Nanoengineering

Nanoengineering

Timothy Sawyer

CLANRYE
INTERNATIONAL
www.clanryeinternational.com

Clanrye International,
750 Third Avenue, 9th Floor,
New York, NY 10017, USA

ISBN: 978-1-64726-133-7

Cataloging-in-Publication Data

Nanoengineering / Timothy Sawyer
        p. cm.
Includes bibliographical references and index.
ISBN: 978-1-64726-133-7
1. Nanotechnology. 2. Nanoscience. 3. Nanostructures. I. Sawyer, Timothy.
T174.7 .N36 2022
620.5--dc23

For information on all Clanrye International publications
visit our website at www.clanryeinternational.com

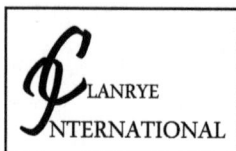

CLANRYE
INTERNATIONAL

TABLE OF CONTENTS

This book aims to help a broader range of students by exploring a wide variety of significant topics related to this discipline. It will help students in achieving a higher level of understanding of the subject and excel in their respective fields. This book would not have been possible without the unwavered support of my senior professors who took out the time to provide me feedback and help me with the process. I would also like to thank my family for their patience and support.

Nanoengineering is the field of engineering which focuses on nanoscale. It studies the development and refinement of nanomaterials. It is considered as a practical application of nanoscience, because it does not emphasize on pure science aspects. Examples of nanomaterials include nanocomposites, nanotubes and quantum dots. It also deals with how various nanomaterials interact to make useful structures, systems, devices and materials. There are various techniques used in nanoengineering. They are molecular self-assembly and scanning tunneling microscope. Scanning tunneling microscope is an instrument, which is helpful in imaging surfaces at atomic levels. It can be used for imaging, as well as manipulating structures of atom. The process of molecular self-assembly is used to create regular patterns of amino acids and custom proteins. Nanoengineering is an upcoming field of science that has undergone rapid development over the past few decades. This book attempts to understand the multiple branches that fall under this discipline and how such concepts have practical applications. This book will serve as a reference to a broad spectrum of readers.

A brief overview of the book contents is provided below:

Chapter – What is Nanoengineering?

The branch of engineering that is concerned with the design, building and use of machines at the nanoscale is called nanoengineering. It deals with nanomaterials and their processes in making useful structures, devices and systems. This is an introductory chapter which will briefly introduce all the significant aspects of nanoenegineering.

Chapter – Understanding Nanotechnology

Nanotechnology refers to the imaging, measuring and modeling of matter on an atomic, molecular and supermolecular scale. Some of the fields that fall under the domain of nanotechnology are molecular nanotechnology, nanorobotics, green nanotechnology, nanoremediation, DNA nanotechnology, etc. All these fields of nanotechnology have been carefully analyzed in this chapter.

Chapter – Nanomaterials

The materials whose size lies between 1 and 1000 nanometers are known as nanomaterials. Some of these are nanorods, nanoshells, nanofibers, silicene, nanosheets and nanomesh. This chapter has been carefully written to provide an easy understanding of these types of nanomaterials.

Chapter – Nanoelectronics

The utilization of nanotechnology in electronic components is referred to as nanoelectronics. Various components of nanoelectronics are nanoradios, nanomotors, nanoionic devices, nanobatteries, nanogenerators, nanosensors and nanowires. The topics elaborated in this chapter will help in gaining a better perspective about these components of nanoelectronics.

Chapter – Nanophotonics

The study of the behavior of light on the nanometer scale and the interaction of nanometer scale objects with light is termed as nanophotonics. The major focus areas of nanophotonics are photonic crystal and surface plasma. This chapter closely examines these key concepts and areas of nanophotonics to provide an extensive understanding of the subject.

Timothy Sawyer

What is Nanoengineering?

The branch of engineering that is concerned with the design, building and use of machines at the nanoscale is called nanoengineering. It deals with nanomaterials and their processes in making useful structures, devices and systems. This is an introductory chapter which will briefly introduce all the significant aspects of nanoenegineering.

Nanoengineering is a branch of engineering that deals with all aspects of the design, building, and use of engines, machines, and structures on the nanoscale. At its core, nanoengineering deals with nanomaterials and how they interact to make useful materials, structures, devices and systems.

Nanoengineering is not exactly a new science, but, rather, an enabling technology with applications in most industries from electronics, to energy, medicine, and biotechnology.

While the term nanoengineering is often used synonymously with the more general term nanotechnology, the former technically focuses more closely on the engineering aspects of the field, as opposed to the broader science and general technology aspects that are encompassed by the latter.

Other closely related terms used in this context are nanofabrication and nanomanufacturing. One possible approach to distinguish between the terms is by using the criterion of economic viability: The connotations of industrial scale and profitability associated with the word manufacturing imply that nanomanufacturing is an economic activity with industrial production facilities with more or less fully automated assembly lines. By contrast, nanofabrication is more of a research activity based on developing new materials and processes – it's more a domain of skilled craftsmen and not of mass production.

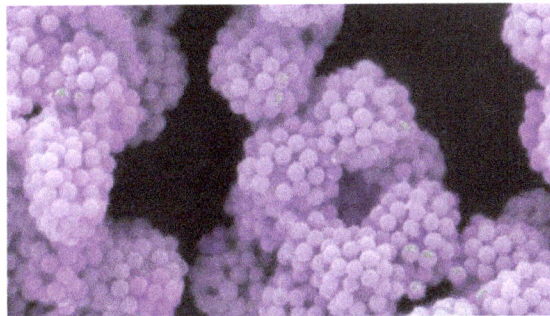

Tiny particles transformed into 'LEGO-like' modular building blocks.

In general, engineering is the branch of science and technology concerned with the design, building, and use of engines, machines, and structures. Correspondingly, but at the scale of atoms and molecules, nanoengineeering exploit the unique properties of nanoscale materials (size and quantum effects) in order to design and manufacture devices and systems that possess entirely new functionality and capabilities.

The properties of materials can be different at the nanoscale for two main reasons:

First, nanomaterials have a relatively larger surface area when compared to the same mass of material produced in a larger form. This can make materials more chemically reactive (in some cases materials that are inert in their larger form are reactive when produced in their nanoscale form), and affect their strength or electrical properties.

Second, quantum effects can begin to dominate the behavior of matter at the nanoscale – particularly at the lower end – affecting the optical, electrical and magnetic behavior of materials. Materials can be produced that are nanoscale in one dimension (for example, nanowires, nanorods and nanotubes), in two dimensions (plate-like shapes like nanocoatings, nanolayers, and graphene) or in all three dimensions (for example, nanoparticles).

Nanoscale objects are difficult to manipulate, as they are too tiny to see directly by eye, far too small to hold, and often have incompatible surfaces for assembling into ordered structures. Therefore the fabrication of complex nanoarchitectures requires sophisticated techniques of nanoscale engineering. To do this, nanoengineers are employing a number of methods to leverage the manipulation of materials on an atomic and molecular scale for (ultimately) industrial purposes.

Among the many challenges that researchers have to overcome in developing nanoengineering techniques and processes, the requirement for extremely precise, nanometer-scale control of positioning and shaping of objects is one of the most vexing.

Nanopatterning and Nanostructuring of Surfaces

There are different ways of fabricating functional nanostructures that you hear most about: top-down and bottom-up methods.

Top-down nanofabrication predominantly is done by photolithography, which is the standard workhorse employed in today's semiconductor industry, and other top-down methods where you start by taking a block of material and remove the bits and pieces you don't want until you get the shape and size you do want.

In contrast to the deterministic nature of top-down processes, bottom-up processes are driven by a combination of thermodynamics and kinetics, which then determines the yield of the desired structure.

Bottom-up processes typically don't need expensive tooling to create nanoscale structures, and scaling to large volumes is potentially straightforward. With the application of the tools of chemical synthesis, quantum dots, plasmonically active particles, carbon nanotubes, metallic nanowires, and multifunctional particles for medical applications have been successfully produced in manufacturing quantities.

Given the vast, and still rising, cost of optical lithography tools, researchers have considered alternative patterning technologies such as electron beam lithography (EBL), and nanoimprint technology (NIL) in order to enable the manufacturing of next-generation integrated circuits, flash memory, and hard disk drives.

Now that the length scales attainable by top-down lithography are approaching that of bottom-up self-assembly found in polymers and small molecules, scientists are increasingly looking at bottom-up patterning technologies based on self-assembly.

The utilization of organic building blocks as self-assembled etch masks is particularly attractive since the lithographic information is contained in the self-assembling material itself, rather than provided in an exposure step.

Architecture is key when designing new materials and DNA is probably the most programmable biomaterial for creating a wide range of rationally designed and functionally enhanced nanostructures.

DNA nanotechnology employs DNA as a programmable building material for self-assembled, nanoscale structures with precisely controlled structures, that can lead to improved surface properties relevant to biosensing, materials science, and cell biology.

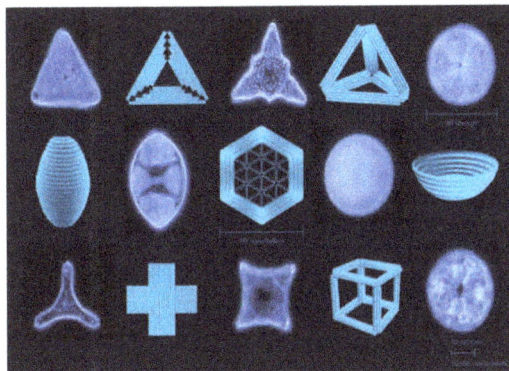

A selection of nanostructures built using DNA origami, alongside naturally-occurring diatoms-single-celled organisms that come in many beautiful and elaborate forms. They are ubiquitous inhabitants of the world's lakes, rivers, and oceans. A scale shows the sizes of the nanostructures and diatoms.

3D Printing

Fabrication of three-dimensional (3D) objects through direct deposition of functional materials – also called additive manufacturing – has been a subject of intense study in the area of macroscale manufacturing for several decades. These 3D printing techniques are reaching a stage where desired products and structures can be made independent of the complexity of their shapes – even bioprinting tissue is now in the realm of the possible.

Applying 3D printing concepts to nanotechnology could bring similar advantages to nanoengineering – speed, less waste, economic viability – than it is expected to bring to manufacturing technologies.

In addition, pre-patterned micro- or nanostructures could be used as substrates, allowing researchers to realize unprecedented manufacturing flexibility, functionality and complexity at the nanoscale.

Already, 3D printing can now be used to print lithium-ion microbatteries the size of a grain of sand. To make the microbatteries, a team based at Harvard University and the University of Illinois at Urbana-Champaign printed precisely interlaced stacks of tiny battery electrodes, each less than the width of a human hair.

This image shows the interlaced stack of electrodes that were printed layer by layer to create the working anode and cathode of a microbattery.

There are various approaches to 3D nanoprinting. In one, researchers use a novel nanoscale additive manufacturing technique termed Nanotribological Printing, which creates structures through tribomechanical and tribochemical surface interactions at the contact between a substrate and an atomic force microscope probe, where material pattern formation is driven by normal and shear contact stresses.

Another technique has been developed exploiting a size-controllable liquid meniscus to engineer 3D printed nanostructures composed entirely of graphene.

For instance, printing three-dimensional objects with incredibly fine details is already possible using a direct laser writing method called two-photon lithography. With this technology, tiny structures on a nanometer scale can be fabricated such as the example of the race car – which measures only 285 µm in length.

Understanding Nanotechnology

Nanotechnology refers to the imaging, measuring and modeling of matter on an atomic, molecular and supermolecular scale. Some of the fields that fall under the domain of nanotechnology are molecular nanotechnology, nanorobotics, green nanotechnology, nanoremediation, DNA nanotechnology, etc. All these fields of nanotechnology have been carefully analyzed in this chapter.

Nanotechnology is the manipulation and manufacture of materials and devices on the scale of atoms or small groups of atoms. The "nanoscale" is typically measured in nanometres, or billionths of a metre, and materials built at this scale often exhibit distinctive physical and chemical properties due to quantum mechanical effects. Although usable devices this small may be decades away, techniques for working at the nanoscale have become essential to electronic engineering, and nanoengineered materials have begun to appear in consumer products. For example, billions of microscopic "nanowhiskers," each about 10 nanometres in length, have been molecularly hooked onto natural and synthetic fibres to impart stain resistance to clothing and other fabrics; zinc oxide nanocrystals have been used to create invisible sunscreens that block ultraviolet light; and silver nanocrystals have been embedded in bandages to kill bacteria and prevent infection.

Possibilities for the future are numerous. Nanotechnology may make it possible to manufacture lighter, stronger, and programmable materials that require less energy to produce than conventional materials, that produce less waste than with conventional manufacturing, and that promise greater fuel efficiency in land transportation, ships, aircraft, and space vehicles. Nanocoatings for both opaque and translucent surfaces may render them resistant to corrosion, scratches, and radiation. Nanoscale electronic, magnetic, and mechanical devices and systems with unprecedented levels of information processing may be fabricated, as may chemical, photochemical, and biological sensors for protection, health care, manufacturing, and the environment; new photoelectric materials that will enable the manufacture of cost-efficient solar-energy panels; and molecular-semiconductor hybrid devices that may become engines for the next revolution in the information age. The potential for improvements in health, safety, quality of life, and conservation of the environment are vast.

At the same time, significant challenges must be overcome for the benefits of nanotechnology to be realized. Scientists must learn how to manipulate and characterize individual atoms and small groups of atoms reliably. New and improved tools are needed to control the properties and structure of materials at the nanoscale; significant improvements in computer simulations of atomic and molecular structures are essential to the understanding of this realm. Next, new tools and approaches are needed for assembling atoms and molecules into nanoscale systems and for the further assembly of small systems into more-complex objects. Furthermore, nanotechnology products must provide not only improved performance but also lower cost. Finally, without integration of nanoscale objects with systems at the micro- and macroscale (that is, from millionths of a metre up to the millimetre scale), it will be very difficult to exploit many of the unique properties found at the nanoscale.

Nanotechnology is highly interdisciplinary, involving physics, chemistry, biology, materials science, and the full range of the engineering disciplines. The word nanotechnology is widely used as shorthand to refer to both the science and the technology of this emerging field. Narrowly defined, nanoscience concerns a basic understanding of physical, chemical, and biological properties on atomic and near-atomic scales. Nanotechnology, narrowly defined, employs controlled manipulation of these properties to create materials and functional systems with unique capabilities.

In contrast to recent engineering efforts, nature developed "nanotechnologies" over billions of years, employing enzymes and catalysts to organize with exquisite precision different kinds of atoms and molecules into complex microscopic structures that make life possible. These natural products are built with great efficiency and have impressive capabilities, such as the power to harvest solar energy, to convert minerals and water into living cells, to store and process massive amounts of data using large arrays of nerve cells, and to replicate perfectly billions of bits of information stored in molecules of deoxyribonucleic acid (DNA).

There are two principal reasons for qualitative differences in material behavior at the nanoscale (traditionally defined as less than 100 nanometres). First, quantum mechanical effects come into play at very small dimensions and lead to new physics and chemistry. Second, a defining feature at the nanoscale is the very large surface-to-volume ratio of these structures. This means that no atom is very far from a surface or interface, and the behavior of atoms at these higher-energy sites have a significant influence on the properties of the material. For example, the reactivity of a metal catalyst particle generally increases appreciably as its size is reduced—macroscopic gold is chemically inert, whereas at nanoscales gold becomes extremely reactive and catalytic and even melts at a lower temperature. Thus, at nanoscale dimensions material properties depend on and change with size, as well as composition and structure.

Using the processes of nanotechnology, basic industrial production may veer dramatically from the course followed by steel plants and chemical factories of the past. Raw materials will come from the atoms of abundant elements—carbon, hydrogen, and silicon—and these will be manipulated into precise configurations to create nanostructured materials that exhibit exactly the right properties for each particular application. For example, carbon atoms can be bonded together in a number of different geometries to create variously a fibre, a tube, a molecular coating, or a wire, all with the superior strength-to-weight ratio of another carbon material—diamond. Additionally, such material processing need not require smokestacks, power-hungry industrial machinery, or intensive human labour. Instead, it may be accomplished either by "growing" new structures through some combination of chemical catalysts and synthetic enzymes or by building them through new techniques based on patterning and self-assembly of nanoscale materials into useful predetermined designs. Nanotechnology ultimately may allow people to fabricate almost any type of material or product allowable under the laws of physics and chemistry. While such possibilities seem remote, even approaching nature's virtuosity in energy-efficient fabrication would be revolutionary.

Even more revolutionary would be the fabrication of nanoscale machines and devices for incorporation into micro- and macroscale systems. Once again, nature has led the way with the fabrication of both linear and rotary molecular motors. These biological machines carry out such tasks as muscle contraction (in organisms ranging from clams to humans) and shuttling little packets of material around within cells while being powered by the recyclable, energy-efficient fuel adenosine triphosphate. Scientists are only beginning to develop the tools to fabricate functioning systems at such small scales, with most advances based on electronic or magnetic information processing and storage systems. The energy-efficient, reconfigurable, and self-repairing aspects of biological systems are just becoming understood.

The potential impact of nanotechnology processes, machines, and products is expected to be far-reaching, affecting nearly every conceivable information technology, energy source, agricultural product, medical device, pharmaceutical, and material used in manufacturing. Meanwhile, the dimensions of electronic circuits on semiconductors continue to shrink, with minimum feature sizes now reaching the nanorealm, under 100 nanometres. Likewise, magnetic memory materials, which form the basis of hard disk drives, have achieved dramatically greater memory density as a result of nanoscale structuring to exploit new magnetic effects at nanodimensions. These latter two areas represent another major trend, the evolution of critical elements of microtechnology into the realm of nanotechnology to enhance performance. They are immense markets driven by the rapid advance of information technology.

Properties at the Nanoscale

At nanoscale dimensions the properties of materials no longer depend solely on composition and structure in the usual sense. Nanomaterials display new phenomena associated with quantized effects and with the preponderance of surfaces and interfaces.

Quantized effects arise in the nanometre regime because the overall dimensions of objects are comparable to the characteristic wavelength for fundamental excitations in materials. For example, electron wave functions in semiconductors are typically on the order of 10 to 100 nanometres. Such excitations include the wavelength of electrons, photons, phonons, and magnons, to name a few. These excitations carry the quanta of energy through materials and thus determine the dynamics of their propagation and transformation from one form to another. When the size of structures is comparable to the quanta themselves, it influences how these excitations move through and interact in the material. Small structures may limit flow, create wave interference effects, and otherwise bring into play quantum mechanical selection rules not apparent at larger dimensions.

Electronic and Photonic Behavior

Quantum mechanical properties for confinement of electrons in one dimension have long been exploited in solid-state electronics. Semiconductor devices are grown with thin layers of differing composition so that electrons (or "holes" in the case of missing electron charges) can be confined in specific regions of the structure (known as quantum wells). Thin layers with larger energy bandgaps can serve as barriers that restrict the flow of charges to certain conditions under which they can "tunnel" through these barriers—the basis of resonant tunneling diodes. Superlattices are periodic structures of repeating wells that set up a new set of selection rules which affect the conditions for charges to flow through the structure. Superlattices have been exploited in cascade lasers to achieve far infrared wavelengths. Modern telecommunications is based on semiconductor lasers that exploit the unique properties of quantum wells to achieve specific wavelengths and high efficiency.

The propagation of photons is altered dramatically when the size and periodicity of the transient structure approach the wavelength of visible light (400 to 800 nanometres). When photons propagate through a periodically varying dielectric constant—for example, semiconductor posts surrounded by air—quantum mechanical rules define and limit the propagation of the photons depending on their energy (wavelength). This new behavior is analogous to the quantum mechanical rules that define the motion of electrons through crystals, giving bandgaps for semiconductors. In one dimension, compound semiconductor superlattices can be grown epitaxially with the alternating layers having different dielectric constants, thus providing highly reflective mirrors for specific wavelengths as determined by the repeat distance of layers in the superlattice. These structures are used to provide "built-in" mirrors for vertical-cavity surface-emitting lasers, which are used in communications applications. In two and three dimensions, periodic structures known as photonic crystals offer additional control over photon propagation.

Photonic crystals are being explored in a variety of materials and periodicities, such as two-dimensional hexagonal arrays of posts fabricated in compound semiconductors or stacked loglike arrays of silicon bars in three dimensions. The dimensions of these structures depend on the wavelength of light being propagated and are typically in the range of a few hundred nanometres for wavelengths in the visible and near infrared. Photonic crystal properties based on nanostructured materials offer the possibility of confining, steering, and separating light by wavelength on unprecedented small scales and of creating new devices such as lasers that require very low currents to initiate lasing (called near-thresholdless lasers). These structures are being extensively investigated as the tools for nanostructuring materials are steadily advancing. Researchers are particularly interested in the infrared wavelengths, where dimensional control is not as stringent as at the shorter visible wavelengths and where optical communications and chemical sensing provide motivation for potential new applications.

Magnetic, Mechanical and Chemical Behavior

Nanoscale materials also have size-dependent magnetic behavior, mechanical properties, and chemical reactivity. At very small sizes (a few nanometres), magnetic nanoclusters have a single magnetic domain, and the strongly coupled magnetic spins on each atom combine to produce a particle with a single "giant" spin. For example, the giant spin of a ferromagnetic iron particle rotates freely at room temperature for diameters below about 16 nanometres, an effect termed superparamagnetism. Mechanical properties of nanostructured materials can reach exceptional strengths. As a specific example, the introduction of two-nanometre aluminum oxide precipitates into thin films of pure nickel results in yield strengths increasing from 0.15 to 5 gigapascals, which is more than twice that for a hard bearing steel. Another example of exceptional mechanical properties at the nanoscale is the carbon nanotube, which exhibits great strength and stiffness along its longitudinal axis.

The preponderance of surfaces is a major reason for the change in behavior of materials at the nanoscale. Since up to half of all the atoms in nanoparticles are surface atoms, properties such as electrical transport are no longer determined by solid-state bulk phenomena. Likewise, the atoms in nanostructures have a higher average energy than atoms in larger structures, because of the large proportion of surface atoms. For example, catalytic materials have a greater chemical activity per atom of exposed surface as the catalyst is reduced in size at the nanoscale. Defects and impurities may be attracted to surfaces and interfaces, and interactions between particles at these small dimensions can depend on the structure and nature of chemical bonding at the surface. Molecular monolayers may be used to change or control surface properties and to mediate the interaction between nanoparticles.

Surfaces and their interactions with molecular structures are basic to all biology. The intersection of nanotechnology and biotechnology offers the possibility of achieving new functions and properties with nanostructured surfaces. In this surface- and interface-dominated regime, biology does an exquisite job of selectively controlling functions through a combination of structure and chemical forces. The transcription of information stored in genes and the selectivity of biochemical reactions based on chemical recognition of complex molecules are examples where interfaces play the key role in establishing nanoscale behavior. Atomic forces and chemical bonds dominate at these dimensions, while macroscopic effects—such as convection, turbulence, and momentum (inertial forces)—are of little consequence.

Nanotechnology Research

Nanomaterials

Properties at the nanoscale, material properties—electrical, optical, magnetic, mechanical, and chemical—depend on their exact dimensions. This opens the way for development of new and improved materials through manipulation of their nanostructure. Hierarchical assemblies of nanoscale-engineered materials into larger structures, or their incorporation into devices, provide the basis for tailoring radically new materials and machines.

Nature's assemblies point the way to improving structural materials. The often-cited abalone seashell provides a beautiful example of how the combination of a hard, brittle inorganic material with nanoscale structuring and a soft, "tough" organic material can produce a strong, durable nanocomposite—basically, these nanocomposites are made of calcium carbonate "bricks" held together by a glycoprotein "glue." New engineered materials are emerging—such as polymer-clay nanocomposites—that are not only strong and tough but also lightweight and easier to recycle than conventional reinforced plastics. Such improvements in structural materials are particularly important for the transportation industry, where reduced weight directly translates into improved fuel economy. Other improvements can increase safety or decrease the impact on the environment of fabrication and recycling. Further advances, such as truly smart materials that signal their impending failure or are even able to self-repair flaws, may be possible with composites of the future.

Sensors are central to almost all modern control systems. For example, multiple sensors are used in automobiles for such diverse tasks as engine management, emission control, security, safety, comfort, vehicle monitoring, and diagnostics. While such traditional applications for physical sensing generally rely on microscale sensing devices, the advent of nanoscale materials and structures has led to new electronic, photonic, and magnetic nanosensors, sometimes known as "smart dust." Because of their small size, nanosensors exhibit unprecedented speed and sensitivity, extending in some cases down to the detection of single molecules. For example, nanowires made of carbon nanotubes, silicon, or other semiconductor materials exhibit exceptional sensitivity to chemical species or biological agents. Electrical current through nanowires can be altered by having molecules attached to their surface that locally perturb their electronic band structure. By means of nanowire surfaces coated with sensor molecules that selectively attach particular species, charge-induced changes in current can be used to detect the presence of those species. This same strategy is adopted for many classes of sensing systems. New types of sensors with ultrahigh sensitivity and specificity will have many applications; for example, sensors that can detect cancerous tumours when they consist of only a few cells would be a very significant advance.

Nanomaterials also make excellent filters for trapping heavy metals and other pollutants from industrial wastewater. One of the greatest potential impacts of nanotechnology on the lives of the majority of people on Earth will be in the area of economical water desalination and purification. Nanomaterials will very likely find important use in fuel cells, bioconversion for energy, bioprocessing of food products, waste remediation, and pollution-control systems.

A recent concern regarding nanoparticles is whether their small sizes and novel properties may pose significant health or environmental risks. In general, ultrafine particles—such as the carbon in photocopier toners or in soot produced by combustion engines and factories—have adverse respiratory and cardiovascular effects on people and animals. Studies are under way to determine if specific nanoscale particles pose higher risks that may require special regulatory restrictions. Of particular concern are potential carcinogenic risks from inhaled particles and the possibility for very small nanoparticles to cross the blood-brain barrier to unknown effect. Nanomaterials currently receiving attention from health officials include carbon nanotubes, buckyballs, and cadmium selenide quantum dots. Studies of the absorption through the skin of titanium oxide nanoparticles (used in sunscreens) are also planned. More far-ranging studies of the toxicity, transport, and overall fate of nanoparticles in ecosystems and the environment have not yet been undertaken. Some early animal studies, involving the introduction of very high levels of nanoparticles which resulted in the rapid death of many of the subjects, are quite controversial.

Biomedicine and Health Care

Drug Delivery

Nanotechnology promises to impact medical treatment in multiple ways. First, advances in nanoscale particle design and fabrication provide new options for drug delivery and drug therapies. More than half of the new drugs developed each year are not water-soluble, which makes their delivery difficult. In the form of nanosized particles, however, these drugs are more readily transported to their destination, and they can be delivered in the conventional form of pills.

More important, nanotechnology may enable drugs to be delivered to precisely the right location in the body and to release drug doses on a predetermined schedule for optimal treatment. The general approach is to attach the drug to a nanosized carrier that will release the medicine in the body over an extended period of time or when specifically triggered to do so. In addition, the surfaces of these nanoscale carriers may be treated to seek out and become localized at a disease site—for example, attaching to cancerous tumours. One type of molecule of special interest for these applications is an organic dendrimer. A dendrimer is a special class of polymeric molecule that weaves in and out from a hollow central region. These spherical "fuzz balls" are about the size of a typical protein but cannot unfold like proteins. Interest in dendrimers derives from the ability to tailor their cavity sizes and chemical properties to hold different therapeutic agents. Researchers hope to design different dendrimers that can swell and release their drug on exposure to specifically recognized molecules that indicate a disease target. This same general approach to nanoparticle-directed drug delivery is being explored for other types of nanoparticles as well.

Another approach involves gold-coated nanoshells whose size can be adjusted to absorb light energy at different wavelengths. In particular, infrared light will pass through several centimetres of body tissue, allowing a delicate and precise heating of such capsules in order to release the therapeutic substance within. Furthermore, antibodies may be attached to the outer gold surface of the shells to cause them to bind specifically to certain tumour cells, thereby reducing the damage to surrounding healthy cells.

Bioassays

A second area of intense study in nanomedicine is that of developing new diagnostic tools. Motivation for this work ranges from fundamental biomedical research at the level of single genes or cells to point-of-care applications for health delivery services. With advances in molecular biology, much diagnostic work now focuses on detecting specific biological "signatures." These analyses are referred to as bioassays. Examples include studies to determine which genes are active in response to a particular disease or drug therapy. A general approach involves attaching fluorescing dye molecules to the target biomolecules in order to reveal their concentration.

Another approach to bioassays uses semiconductor nanoparticles, such as cadmium selenide, which emit light of a specific wavelength depending on their size. Different-size particles can be tagged to different receptors so that a wider variety of distinct colour tags are available than can be distinguished for dye molecules. The degradation in fluorescence with repeated excitation for dyes is avoided. Furthermore, various-size particles can be encapsulated in latex beads and their resulting wavelengths read like a bar code. This approach, while still in the exploratory stage, would allow for an enormous number of distinct labels for bioassays.

Approaches that do not involve optical detection techniques are also being explored with nanoparticles. For example, magnetic nanoparticles can be attached to antibodies that in turn recognize and attach to specific biomolecules. The magnetic particles then act as tags and "handlebars" through which magnetic fields can be used for mixing, extracting, or identifying the attached biomolecules within microlitre- or nanolitre-sized samples. For example, magnetic nanoparticles stay magnetized as a single domain for a significant period, which enables them to be aligned and detected in a magnetic field. In particular, attached antibody–magnetic-nanoparticle combinations rotate slowly and give a distinctive magnetic signal. In contrast, magnetically tagged antibodies that are not attached to the biological material being detected rotate more rapidly and so do not give the same distinctive signal.

Microfluidic systems, or "labs-on-chips," have been developed for biochemical assays of minuscule samples. Typically cramming numerous electronic and mechanical components into a portable unit no larger than a credit card, they are especially useful for conducting rapid analysis in the field. While these microfluidic systems primarily operate at the microscale (that is, millionths of a metre), nanotechnology has contributed new concepts and will likely play an increasing role in the future. For example, separation of DNA is sensitive to entropic effects, such as the entropy required to unfold DNA of a given length. A new approach to separating DNA could take advantage of its passage through a nanoscale array of posts or channels such that DNA molecules of different lengths would uncoil at different rates.

Another nanotechnology variation on bioassays is to attach one half of the single-stranded complementary DNA segment for the genetic sequence to be detected to one set of gold particles and the other half to a second set of gold particles. When the material of interest is present in a solution, the two attachments cause the gold balls to agglomerate, providing a large change in optical properties that can be seen in the colour of the solution. If both halves of the sequence do not match, no agglomeration will occur and no change will be observed.

Other researchers have focused on detecting signal changes as nanometre-wide DNA strands are threaded through a nanoscale pore. Early studies used pores punched in membranes by viruses; artificially fabricated nanopores are also being tested. By applying an electric potential across the membrane in a liquid cell to pull the DNA through, changes in ion current can be measured as different repeating base units of the molecule pass through the pores. Nanotechnology-enabled advances in the entire area of bioassays will clearly impact health care in many ways, from early detection, rapid clinical analysis, and home monitoring to new understanding of molecular biology and genetic-based treatments for fighting disease.

Assistive Devices and Tissue Engineering

Another biomedical application of nanotechnology involves assistive devices for people who have lost or lack certain natural capabilities. For example, researchers hope to design retinal implants for vision-impaired individuals. The concept is to implant chips with photodetector arrays to transmit signals from the retina to the brain via the optic nerve. Meaningful spatial information, even if only at a rudimentary level, would be of great assistance to the blind. Such research illustrates the tremendous challenge of designing hybrid systems that work at the interface between inorganic devices and biological systems.

Closely related research involves implanting nanoscale neural probes in brain tissue to activate and control motor functions. This requires effective and stable "wiring" of many electrodes to neurons. It is exciting because of the possibility of recovery of control for motor-impaired individuals. Studies employing neural stimulation of damaged spinal cords by electrical signals have demonstrated the return of some locomotion. Researchers are also seeking ways to assist in the regeneration and healing of bone, skin, and cartilage—for example, developing synthetic biocompatible or biodegradable structures with nanosized voids that would serve as templates for regenerating specific tissue while delivering chemicals to assist in the repair process. At a more sophisticated level, researchers hope to someday build nanoscale or microscale machines that can repair, assist, or replace more-complex organs.

Information Technology

Semiconductor experts agree that the ongoing shrinkage in "conventional" electronic devices will inevitably reach fundamental limits due to quantum effects such as "tunneling," in which electrons jump out of their prescribed circuit path and create atomic-scale interference between devices. At that point, radical new approaches to data storage and information processing will be required for further advances. For example, radically new systems have been imagined that are based on quantum computing or biomolecular computing.

Molecular Electronics

The use of molecules for electronic devices was suggested by Mark Ratner of Northwestern University and Avi Aviram of IBM as early as the 1970s, but proper nanotechnology tools did not become available until the turn of the 21st century. Wiring up molecules some half a nanometre wide and a few nanometres long remains a major challenge, and an understanding of electrical transport through single molecules is only beginning to emerge. A number of groups have been able to demonstrate molecular switches, for example, that could conceivably be used in computer memory or logic arrays. Current areas of research include mechanisms to guide the selection of molecules, architectures for assembling molecules into nanoscale gates, and three-terminal molecules for transistor-like behavior. More-radical approaches include DNA computing, where single-stranded DNA on a silicon chip would encode all possible variable values and complementary strand interactions would be used for a parallel processing approach to finding solutions. An area related to molecular electronics is that of organic thin-film transistors and light emitters, which promise new applications such as video displays that can be rolled out like wallpaper and flexible electronic newspapers.

Nanotubes and Nanowires

Carbon nanotubes have remarkable electronic, mechanical, and chemical properties. Depending on their specific diameter and the bonding arrangement of their carbon atoms, nanotubes exhibit either metallic or semiconducting behavior. Electrical conduction within a perfect nanotube is ballistic (negligible scattering), with low thermal dissipation. As a result, a wire made from a nanotube, or a nanowire, can carry much more current than an ordinary metal wire of comparable size. At 1.4 nanometres in diameter, nanotubes are about a hundred times smaller than the gate width of silicon semiconductor devices. In addition to nanowires for conduction, transistors, diodes, and simple logic circuits have been demonstrated by combining metallic and semiconductor carbon nanotubes. Similarly, silicon nanowires have been used to build experimental devices, such as field-effect transistors, bipolar transistors, inverters, light-emitting diodes, sensors, and even simple memory. A major challenge for nanowire circuits, as for molecular electronics, is connecting and integrating these devices into a workable high-density architecture. Ideally, the structure would be grown and assembled in place. Crossbar architectures that combine the function of wires and devices are of particular interest.

Single-electron Transistors

At nanoscale dimensions the energy required to add one additional electron to a "small island" (isolated physical region)—for example, through a tunneling barrier—becomes significant. This change in energy provides the basis for devising single-electron transistors. At low temperatures, where thermal fluctuations are small, various single-electron-device nanostructures are readily achievable, and extensive research has been carried out for structures with confined electron flow. However, room-temperature applications will require that sizes be reduced significantly, to the one-nanometre range, to achieve stable operation. For large-scale application with millions of devices, as found in current integrated circuits, the need for structures with very uniform size to maintain uniform device characteristics presents a significant challenge. Also, in this and many new nanodevices being explored, the lack of gain is a serious drawback limiting implementation in large-scale electronic circuits.

Spintronics

Spintronics refers to electronic devices that perform logic operations based on not just the electrical charge of carriers but also their spin. For example, information could be transported or stored through the spin-up or spin-down states of electrons. This is a new area of research, and issues include the injection of spin-polarized carriers, their transport, and their detection. The role of nanoscale structure and electronic properties of the ferromagnetic-semiconductor interface on the spin injection process, the growth of new ferromagnetic semiconductors with nanoscale control, and the possible use of nanostructured features to manipulate spin are all of interest.

Information Storage

Current approaches to information storage and retrieval include high-density, high-speed, solid-state electronic memories, as well as slower (but generally more spacious) magnetic and optical discs. As the minimum feature size for electronic processing approaches 100 nanometres, nanotechnology provides ways to decrease further the bit size of the stored information, thus increasing density and reducing interconnection distances for obtaining still-higher speeds. For example, the basis of the current generation of magnetic disks is the giant magnetoresistance effect. A magnetic read/write head stores bits of information by setting the direction of the magnetic field in nanometre-thick metallic layers that alternate between ferromagnetic and nonferromagnetic. Differences in spin-dependent scattering of electrons at the interface layers lead to resistance differences that can be read by the magnetic head. Mechanical properties, particularly tribology (friction and wear of moving surfaces), also play an important role in magnetic hard disk drives, since magnetic heads float only about 10 nanometres above spinning magnetic disks.

Another approach to information storage that is dependent on designing nanometre-thick magnetic layers is under commercial development. Known as magnetic random access memory (MRAM), a line of electrically switchable magnetic material is separated from a permanently magnetized layer by a nanoscale nonmagnetic interlayer. A resistance change that depends on the relative alignment of the fields is read electrically from a large array of wires through cross lines. MRAM will require a relatively small evolution from conventional semiconductor manufacturing, and it has the added benefit of producing nonvolatile memory (no power or batteries are needed to maintain stored memory states).

Still at an exploratory stage, studies of electrical conduction through molecules have generated interest in their possible use as memory. While still very speculative, molecular and nanowire approaches to memory are intriguing because of the small volume in which the bits of memory are stored and the effectiveness with which biological systems store large amounts of information.

Communications

Nanoscale structuring of optical devices, such as vertical-cavity surface-emitting lasers (VCSELs), quantum dot lasers, and photonic crystal materials, is leading to additional advances in communications technology.

VCSELs have nanoscale layers of compound semiconductors epitaxially grown into their structure—alternating dielectric layers as mirrors and quantum wells. Quantum wells allow the charge carriers to be confined in well-defined regions and provide the energy conversion into light at desired wavelengths. They are placed in the laser's cavity to confine carriers at the nodes of a standing wave and to tailor the band structure for more efficient radiative recombination. One-dimensional nanotechnology techniques involving precise growth of very thin epitaxial semiconductor layers were developed during the 1990s. Such nanostructuring has enhanced the efficiency of VCSELs and reduced the current required for lasing to start (called the threshold current). Because of improving performance and their compatibility with planar manufacturing technology, VCSELs are fast becoming a preferred laser source in a variety of communications applications.

More recently, the introduction of quantum dots (regions so small that they can be given a single electric charge) into semiconductor lasers has been investigated and found to give additional benefits—both further reductions in threshold current and narrower line widths. Quantum dots further confine the optical emission modes within a very narrow spectrum and give the lowest threshold current densities for lasing achieved to date in VCSELs. The quantum dots are introduced into the laser during the growth of strained layers, by a process called Stransky-Krastanov growth. They arise because of the lattice mismatch stress and surface tension of the growing film. Improvements in ways to control precisely the resulting quantum dots to a more uniform single size are still being sought.

Photonic crystals provide a new means to control the steering and manipulation of photons based on periodic dielectric lattices with repeat dimensions on the order of the wavelength of light. These materials can have very exotic properties, such as not allowing light within certain wavelengths to be propagated in a material based on the particular periodic structure. Photonic lattices can act as perfect wavelength-selective mirrors to reflect back incident light from all orientations. They provide the basis for optical switching, steering, and wavelength separation on unprecedented small scales. The periodic structures required for these artificial crystals can be configured as both two- and three-dimensional lattices. Optical sources, switches, and routers are being considered, with two-dimensional planar geometries receiving the most attention, because of their greater ease of fabrication.

Another potentially important communications application for nanotechnology is microelectromechanical systems (MEMS), devices sized at the micrometre level (millionths of a metre). MEMS are currently poised to have a major impact on communications via optical switching. In the future, electromechanical devices may shrink to nanodimensions to take advantage of the higher frequencies of mechanical vibration at smaller masses. The natural (resonant) frequency of vibration for small mechanical beams increases as their size decreases, so that little power is needed to drive them as oscillators. Their efficiency is rated by a quality factor, known as Q, which is a ratio of the energy stored per cycle versus the energy dissipated per cycle. The higher the Q, the more precise the absolute frequency of an oscillator. The Q is very high for micro- and nanoscale mechanical oscillators, and these devices can reach very high frequencies (up to microwave frequencies), making them potential low-power replacements for electronic-based oscillators and filters.

Mechanical oscillators have been made from silicon at dimensions of 10 × 100 nanometres, where more than 10 percent of the atoms are less than one atomic distance from the surface. While highly homogeneous materials can be made at these dimensions—for example, single-crystal silicon bars—surfaces play an increasing role at nanoscales, and energy losses increase, presumably because of surface defects and molecular species absorbed on surfaces.

It is possible to envision even higher frequencies, in what might be viewed as the ultimate in nanomechanical systems, by moving from nanomachined structures to molecular systems. As an example, multiwalled carbon nanotubes are being explored for their mechanical properties. When the ends of the outer nanotube are removed, the inner tube may be pulled partway out from the outer tube where van der Waals forces between the two tubes will supply a restoring force. The inner tube can thus oscillate, sliding back and forth inside the outer tube. The resonant frequency of oscillation for such structures is predicted to be above one gigahertz (one billion cycles per second). It is unknown whether connecting such systems to the macro world and protecting them from surface effects will ever be practical.

Nanofabrication

Two very different paths are pursued. One is a top-down strategy of miniaturizing current technologies, while the other is a bottom-up strategy of building ever-more-complex molecular devices atom by atom. Top-down approaches are good for producing structures with long-range order and for making macroscopic connections, while bottom-up approaches are best suited for assembly and establishing short-range order at nanoscale dimensions. The integration of top-down and bottom-up techniques is expected to eventually provide the best combination of tools for nanofabrication. Nanotechnology requires new tools for fabrication and measurement.

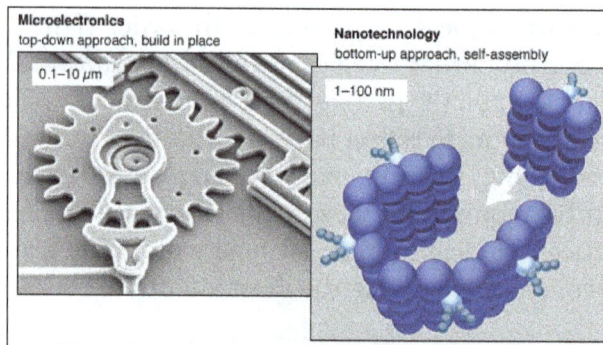

Top-down approaches have been developed for building structures at the scale of the micrometre (μm). Bottom-up techniques have also been developed for assembling small groups of atoms or molecules at the scale of nanometres (nm). The remaining task is to combine these approaches in order to create extended structures at the nanoscale.

Top-down Approach

The most common top-down approach to fabrication involves lithographic patterning techniques using short-wavelength optical sources. A key advantage of the top-down approach—as developed in the fabrication of integrated circuits—is that the parts are both patterned and built in place, so that no assembly step is needed. Optical lithography is a relatively mature field because of the high degree of refinement in microelectronic chip manufacturing, with current short-wavelength optical lithography techniques reaching dimensions just below 100 nanometres (the traditional threshold definition of the nanoscale). Shorter-wavelength sources, such as extreme ultraviolet and X-ray, are being developed to allow lithographic printing techniques to reach dimensions from 10 to 100 nanometres. Scanning beam techniques such as electron-beam lithography provide patterns down to about 20 nanometres. Here the pattern is written by sweeping a finely focused electron beam across the surface. Focused ion beams are also used for direct processing and patterning of wafers, although with somewhat less resolution than in electron-beam lithography. Still-smaller features are obtained by using scanning probes to deposit or remove thin layers.

Mechanical printing techniques—nanoscale imprinting, stamping, and molding—have been extended to the surprisingly small dimensions of about 20 to 40 nanometres. The details of these techniques vary, but they are all based on making a master "stamp" by a high-resolution technique such as electron-beam lithography and then applying this stamp, or subsequent generations of it, to a surface to create the pattern. In one variation a stamp's surface is coated with a very thin layer of material (the "ink") that can then be deposited ("inked") directly onto the surface to reproduce the stamp's pattern. For example, the controlled patterning of a molecular monolayer on a surface can be achieved by stamping an ink of thiol functionalized organic molecules directly onto a gold-coated surface (molecules that contain a sulfur end group, called a thiol, bond strongly to gold). In another approach the stamp is used mechanically to press the pattern into a thin layer of material. This surface layer is typically a polymeric material that has been made pliable for the molding process by being heated during the stamping procedure. Plasma etching can then be used to remove the thin layer of the masking material under the stamped regions; any residual polymer is thus removed, and a nanoscale lithographic pattern is left on the surface. Still another variation is to make the relief pattern out of photoresist on a silicon wafer by optical or electron-beam lithography and then pour a liquid precursor—for example, polydimethylsiloxane, a form of silicone— over the pattern and then cure it. The result is a rubbery solid that can be peeled off and used as a stamp. These stamps can be inked and printed as described above, or they can be pressed to the surface and a liquid polymer allowed to flow into the raised regions of the mask by capillary action and cured in place. A distinction for this latter approach is that the stamp is flexible and can thus be used to print nanoscale features on curved surfaces.

These nanoscale printing techniques offer several advantages beyond the ability to use a wider variety of materials with curved surfaces. In particular, such approaches can be carried out in ordinary laboratories with far-less-expensive equipment than that needed for conventional submicron lithography. The challenge for all top-down techniques is that, while they work well at the microscale (at millionths of a metre), it becomes increasingly difficult to apply them at nanoscale dimensions. A second disadvantage is that they involve planar techniques, which means that structures are created by the addition and subtraction of patterned layers (deposition and etching), so arbitrary three-dimensional objects are difficult to construct.

Bottom-up Approach

Bottom-up, or self-assembly, approaches to nanofabrication use chemical or physical forces operating at the nanoscale to assemble basic units into larger structures. As component size decreases in nanofabrication, bottom-up approaches provide an increasingly important complement to top-down techniques. Inspiration for bottom-up approaches comes from biological systems, where nature has harnessed chemical forces to create essentially all the structures needed by life. Researchers hope to replicate nature's ability to produce small clusters of specific atoms, which can then self-assemble into more-elaborate structures.

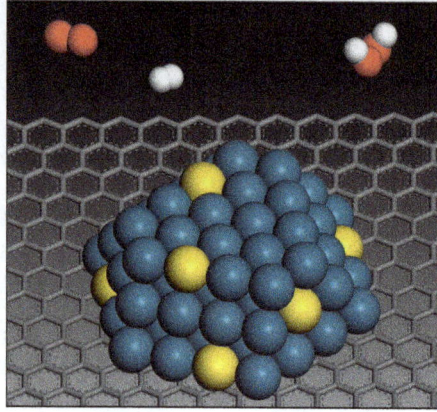

Nanoparticles: hydrogen peroxide.

In figure nanoparticles of a gold-palladium (yellow-blue) alloy supported on acid-treated carbon (gray) directly catalyzing hydrogen peroxide formation from hydrogen (white) and oxygen (red) while shutting off unwanted hydrogen peroxide decomposition.

A number of bottom-up approaches have been developed for producing nanoparticles, ranging from condensation of atomic vapours on surfaces to coalescence of atoms in liquids. For example, liquid-phase techniques based on inverse micelles (globules of lipid molecules floating in a non-aqueous solution in which their polar, or hydrophilic, ends point inward to form a hollow core, as shown in the figure) have been developed to produce size-selected nanoparticles of semiconductor, magnetic, and other materials. An example of self-assembly that achieves a limited degree of control over both formation and organization is the growth of quantum dots. Indium gallium arsenide (InGaAs) dots can be formed by growing thin layers of InGaAs on GaAs in such a manner that repulsive forces caused by compressive strain in the InGaAs layer results in the formation of isolated quantum dots. After the growth of multiple layer pairs, a fairly uniform spacing of the dots can be achieved. Another example of self-assembly of an intricate structure is the formation of carbon nanotubes under the right set of chemical and temperature conditions.

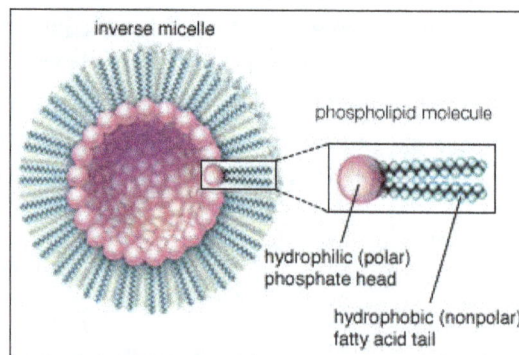

Phospholipid molecules, composed of fatty acid "tails" and a phosphate "head," form an inverse micelle in a nonaqueous solution. The phosphate group converts one end of the lipid molecule into a polar, or hydrophilic, group, leaving the preferentially attracted nonpolar, or hydrophobic, end of the molecule to react with the nonaqueous solution.

DNA-assisted assembly may provide a method to integrate hybrid heterogeneous parts into a single device. Biology does this very well, combining self-assembly and self-organization in fluidic environments where weaker electrochemical forces play a significant role. By using DNA-like recognition, molecules on surfaces may be able to direct attachments between objects in fluids. In this approach, polymers made with complementary DNA strands would be used as intelligent "adhesive tape," attaching between polymers only when the right pairing is present. Such assembly might be combined with electrical fields to assist in locating the attachment sites and then be followed by more-permanent attachment approaches, such as electrodeposition and metallization. There are several advantages of DNA-assisted approaches: DNA molecules can be sequenced and replicated in large quantities, DNA sequences act as codes that can be used to recognize complementary DNA strands, hybridized DNA strands form strong bonds to their complementary sequence, and DNA strands can be attached to different devices as labels. These properties are being explored for ways to self-assemble molecules into nanoscale units. For example, sequences of DNA have been fabricated that adhere only to particular crystal faces of compound semiconductors, providing a basis for self-assembly. By having the correct complementary sequences at the other end of the DNA molecule, certain faces of small semiconductor building blocks can be made that adhere to or repel each other. For example, thiol groups at the end of molecules cause them to attach to gold surfaces, while carboxyl groups can be used for attachment to silica surfaces. Directed assembly is an increasingly important variation of self-assembly where, in quasi-equilibrium environments, parts are moved mechanically, electrically, or magnetically and are placed precisely where they are intended to go.

ENERGY APPLICATIONS OF NANOTECHNOLOGY

Over the past few decades, the fields of science and engineering have been seeking to develop new and improved types of energy technologies that have the capability of improving life all over the world. In order to make the next leap forward from the current generation of technology, scientists and engineers have been developing energy applications of nanotechnology. Nanotechnology, a new field in science, is any technology that contains components smaller than 100 nanometers. For scale, a single virus particle is about 100 nanometers in width.

An important subfield of nanotechnology related to energy is nanofabrication. Nanofabrication is the process of designing and creating devices on the nanoscale. Creating devices smaller than 100 nanometers opens many doors for the development of new ways to capture, store, and transfer energy. The inherent level of control that nanofabrication could give scientists and engineers would be critical in providing the capability of solving many of the problems that the world is facing today related to the current generation of energy technologies.

People in the fields of science and engineering have already begun developing ways of utilizing nanotechnology for the development of consumer products. Benefits already observed from the design of these products are an increased efficiency of lighting and heating, increased electrical storage capacity, and a decrease in the amount of pollution from the use of energy. Benefits such as these make the investment of capital in the research and development of nanotechnology a top priority.

Consumer Products

Recently, previously established and entirely new companies such as BetaBatt, Inc. and Oxane Materials are focusing on nanomaterials as a way to develop and improve upon older methods for the capture, transfer, and storage of energy for the development of consumer products.

ConsERV, a product developed by the Dais Analytic Corporation, uses nanoscale polymer membranes to increase the efficiency of heating and cooling systems and has already proven to be a lucrative design. The polymer membrane was specifically configured for this application by selectively engineering the size of the pores in the membrane to prevent air from passing, while allowing moisture to pass through the membrane. ConsERV's value is demonstrated in the form of an energy recovery a device which pretreats the incoming fresh air to a building using the energy found in the exhaust air steam using no moving parts to lower the energy and carbon footprint of existing forms of heating and cooling equipment Polymer membranes can be designed to selectively allow particles of one size and shape to pass through while preventing other. This makes for a powerful tool that can be used in all markets - consumer, commercial, industrial, and government products from biological weapons protection to industrial chemical separations. Dais's near term uses of this 'family' of selectively engineered nanotechnology materials, aside from ConsERV, include (a.) a completely new cooling cycle capable of replacing the refrigerant based cooling cycle the world has known for the past 100 plus years. This product, under development, is named NanoAir. NanoAir uses only water and this selectively engineered membrane material to cool (or heat) and dehumidify (or humidify) air. There are no fluorocarbon producing gasses used, and the energy required to cool a space drops as thermodynamics does the actual cooling. The company was awarded an Advanced Research Program Administration - Energy award in 2010, and a United States Department of Defense (DoD) grant in 2011 both designed to accelerate this newer, energy efficient technology closer to commercialization, and (b.) a novel way to clean most all contaminated forms of water called NanoClear. By using the selectivity of this hermetic, engineered composite material it can transfer only a water molecule from one face of the membrane to the other leaving behind the contaminants. It should also be noted Dais received a US Patent (Patent Number 7,990,679) in October 2011 titled "Nanoparticle Ultracapacitor". This patented item again uses the selectively engineered material to create an energy storage mechanism projected to have performance and cost advantages over existing storage technologies. The company has used this patent's concepts to create a functional energy storage prototype device named NanoCap. NanoCap is a form of ultra-capacitor potentially useful to power a broad range of applications including most forms of transportation, energy storage (especially useful as a storage media for renewable energy technologies), telecommunication infrastructure, transistor gate dielectrics, and consumer battery applications (cell phones, computers, etc.).

A New York-based company called Applied NanoWorks, Inc. has been developing a consumer product that utilizes LED technology to generate light. Light-emitting diodes or LEDs, use only about 10% of the energy that a typical incandescent or fluorescent light bulb uses and typically last much longer, which makes them a viable alternative to traditional light bulbs. While LEDs have been around for decades, this company and others like it have been developing a special variant of LED called the white LED. White LEDs consist of semi-conducting organic layers that are only about 100 nanometers in distance from each other and are placed between two electrodes, which create an anode, and a cathode. When voltage is applied to the system, light is generated when electricity passes through the two organic layers. This is called electroluminescence. The semiconductor properties of the organic layers are what allow for the minimal amount of energy necessary to generate light. In traditional light bulbs, a metal filament is used to generate light when electricity is run through the filament. Using metal generates a great deal of heat and therefore lowers efficiency.

Research for longer lasting batteries has been an ongoing process for years. Researchers have now begun to utilize nanotechnology for battery technology. mPhase Technologies in conglomeration with Rutgers University and Bell Laboratories have utilized nanomaterials to alter the wetting behavior of the surface where the liquid in the battery lies to spread the liquid droplets over a greater area on the surface and therefore have greater control over the movement of the droplets. This gives more control to the designer of the battery. This control prevents reactions in the battery by separating the electrolytic liquid from the anode and the cathode when the battery is not in use and joining them when the battery is in need of use.

Thermal applications also are a future applications of nanothechonlogy creating low cost system of heating, ventilation, and air conditioning, changing molecular structure for better management of temperature.

Reduction of Energy Consumption

A reduction of energy consumption can be reached by better insulation systems, by the use of more efficient lighting or combustion systems, and by use of lighter and stronger materials in the transportation sector. Currently used light bulbs only convert approximately 5% of the electrical energy into light. Nanotechnological approaches like or quantum caged atoms (QCAs) could lead to a strong reduction of energy consumption for illumination.

Increasing the Efficiency of Energy Production

Today's best solar cells have layers of several different semiconductors stacked together to absorb light at different energies but they still only manage to use 40 percent of the Sun's energy. Commercially available solar cells have much lower efficiencies (15-20%). Nanostructuring has been used to improve the efficiencies of established photovoltaic technologies, for example by improving current collection in amorphous silicon devices, plasmonic enhancement in dye-sensitized solar cells, and improved light trapping in crystalline silicon. Furthermore, nanotechnology could help increase the efficiency of light conversion by using nanostructures with a continuum of bandgaps, or by controlling the directivity and photon escape probability of photovoltaic devices.

The degree of efficiency of the internal combustion engine is about 30-40% at present. Nanotechnology could improve combustion by designing specific catalysts with maximized surface area. In 2005, scientists at the University of Toronto developed a spray-on nanoparticle substance that, when applied to a surface, instantly transforms it into a solar collector.

Nuclear Accident Cleanup and Waste Storage

Nanomaterials deployed by swarm robotics may be helpful for decontaminating a site of a nuclear accident which poses hazards to humans because of high levels of radiation and radioactive particles. Hot nuclear compounds such as corium or melting fuel rods may be contained in "bubbles" made from nanomaterials that are designed to isolate the harmful effects of nuclear activity occurring inside of them from the outside environment that organisms inhabit.

Economic Benefits

The relatively recent shift toward using nanotechnology with respect to the capture, transfer, and storage of energy has and will continue to have many positive economic impacts on society. The control of materials that nanotechnology offers to scientists and engineers of consumer products is one of the most important aspects of nanotechnology. This allows for an improved efficiency of products across the board.

A major issue with current energy generation is the loss of efficiency from the generation of heat as a by-product of the process. A common example of this is the heat generated by the internal combustion engine. The internal combustion engine loses about 64% of the energy from gasoline as heat and an improvement of this alone could have a significant economic impact. However, improving the internal combustion engine in this respect has proven to be extremely difficult without sacrificing performance. Improving the efficiency of fuel cells through the use of nanotechnology appears to be more plausible by using molecularly tailored catalysts, polymer membranes, and improved fuel storage.

In order for a fuel cell to operate, particularly of the hydrogen variant, a noble-metal catalyst (usually platinum, which is very expensive) is needed to separate the electrons from the protons of the hydrogen atoms. However, catalysts of this type are extremely sensitive to carbon monoxide reactions. In order to combat this, alcohols or hydrocarbons compounds are used to lower the carbon monoxide concentration in the system. This adds an additional cost to the device. Using nanotechnology, catalysts can be designed through nanofabrication that are much more resistant to carbon monoxide reactions, which improves the efficiency of the process and may be designed with cheaper materials to additionally lower costs.

Fuel cells that are currently designed for transportation need rapid start-up periods for the practicality of consumer use. This process puts a lot of strain on the traditional polymer electrolyte membranes, which decreases the life of the membrane requiring frequent replacement. Using nanotechnology, engineers have the ability to create a much more durable polymer membrane, which addresses this problem. Nanoscale polymer membranes are also much more efficient in ionic conductivity. This improves the efficiency of the system and decreases the time between replacements, which lowers costs.

Another problem with contemporary fuel cells is the storage of the fuel. In the case of hydrogen fuel cells, storing the hydrogen in gaseous rather than liquid form improves the efficiency by 5%. However, the materials that we currently have available to us significantly limit fuel storage due to low stress tolerance and costs. Scientists have come up with an answer to this by using a nanoporous styrene material (which is a relatively inexpensive material) that when super-cooled to around -196 °C, naturally holds on to hydrogen atoms and when heated again releases the hydrogen for use.

Capacitors

For decades, scientists and engineers have been attempting to make computers smaller and more efficient. A crucial component of computers are capacitors. A capacitor is a device that is made of a pair of electrodes separated by an insulator that each stores an opposite charge. A capacitor stores a charge when it is removed from the circuit that it is connected to; the charge is released when it is replaced back into the circuit. Capacitors have an advantage over batteries in that they release their charge much more quickly than a battery.

Traditional or foil capacitors are composed of thin metal conducting plates separated by an electrical insulator, which are then stacked or rolled and placed in a casing. The problem with a traditional capacitor such as this is that they limit how small an engineer can design a computer. Scientists and engineers have since turned to nanotechnology for a solution to the problem.

Using nanotechnology, researchers developed what they call "ultracapacitors." An ultracapacitor is a capacitor that contains nanocomponents. Ultracapacitors are being researched heavily because of their high density interior, compact size, reliability, and high capacitance. This decrease in size makes it increasingly possible to develop much smaller circuits and computers. Ultracapacitors also have the capability to supplement batteries in hybrid vehicles by providing a large amount of energy during peak acceleration and allowing the battery to supply energy over longer periods of time, such as during a constant driving speed. This could decrease the size and weight of the large batteries needed in hybrid vehicles as well as take additional stress off the battery. However, the combination of ultracapacitors and a battery is not cost effective due to the need of additional DC/DC electronics to coordinate the two.

Nanoporous carbon aerogel is one type of material that is being utilized for the design of ultracapacitors. These aerogels have a very large interior surface area and can have its properties altered by changing the pore diameter and distribution along with adding nanosized alkali metals to alter its conductivity.

Carbon nanotubes are another possible material for use in an ultracapacitor. Carbon nanotubes are created by vaporizing carbon and allowing it to condense on a surface. When the carbon condenses, it forms a nanosized tube composed of carbon atoms. This tube has a high surface area, which increases the amount of charge that can be stored. The low reliability and high cost of using carbon nanotubes for ultracapacitors is currently an issue of research.

In a study concerning ultracapacitors or supercapacitors, researchers at the Sungkyunkwan University in the Republic of Korea explored the possibility of increasing the capacitance of electrodes through the addition of fluorine atoms to the walls of carbon nanotubes. As briefly mentioned before, carbon nanotubes are an increasing form of capacitors due to their superb chemical stability, high conductivity, light mass, and their large surface area. These researchers fluorinated single-walled carbon nanotubes (SWCNTs) at high temperatures to bind fluorine atoms to the walls. The attached fluorine atoms changed the non-polar nanotubes to become polar molecules. This can be attributed to the charge transfer from the fluorine. This created dipole-dipole layers along the carbon nanotube walls. Testing of these fluorinated SWCNTs against normal state SWCNTs showed a difference in capacitance. It was determined that the fluorinated SWCNTs are advantageous in fabricating electrodes for capacitors and improve the wettability with aqueous electrolytes, which promotes the overall performance of supercapacitors. While this study brought to knowledge a more efficient example of capacitors, little is known about this new supercapacitor, large scale synthesis is lacking and is necessary for any massive production, and preparation conditions are quite tedious in achieving the final product.

Theory of Capacitance

Understanding the concept of capacitance can be helpful in understanding why nanotechnology is such a powerful tool for the design of higher energy storing capacitors. A capacitor's capacitance (C) or amount of energy stored is equal to the amount of charge (Q) stored on each plate divided by the voltage (V) between the plates. Another representation of capacitance is that capacitance (C) is approximately equal to the permittivity (ε) of the dielectric times the area (A) of the plates divided by the distance (d) between them. Therefore, capacitance is proportional to the surface area of the conducting plate and inversely proportional to the distance between the plates.

Using carbon nanotubes as an example, a property of carbon nanotubes is that they have a very high surface area to store a charge. Using the above proportionality that capacitance (C) is proportional to the surface area (A) of the conducting plate; it becomes obvious that using nanoscaled materials with high surface area would be great for increasing capacitance. The other proportionality described above is that capacitance (C) is inversely proportional to the distance (d) between the plates. Using nanoscaled plates such as carbon nanotubes with nanofabrication techniques, gives the capability of decreasing the space between plates which again increases capacitance.

MOLECULAR NANOTECHNOLOGY

Molecular nanotechnology (MNT) is a technology based on the ability to build structures to complex, atomic specifications by means of mechanosynthesis. This is distinct from nanoscale materials. Based on Richard Feynman's vision of miniature factories using nanomachines to build complex products (including additional nanomachines), this advanced form of nanotechnology (or *molecular manufacturing*) would make use of positionally-controlled mechanosynthesis guided by molecular machine systems. MNT would involve combining physical principles demonstrated by biophysics, chemistry, other nanotechnologies, and the molecular machinery of life with the systems engineering principles found in modern macroscale factories.

While conventional chemistry uses inexact processes obtaining inexact results, and biology exploits inexact processes to obtain definitive results, molecular nanotechnology would employ original definitive processes to obtain definitive results. The desire in molecular nanotechnology would be to balance molecular reactions in positionally-controlled locations and orientations to obtain desired chemical reactions, and then to build systems by further assembling the products of these reactions.

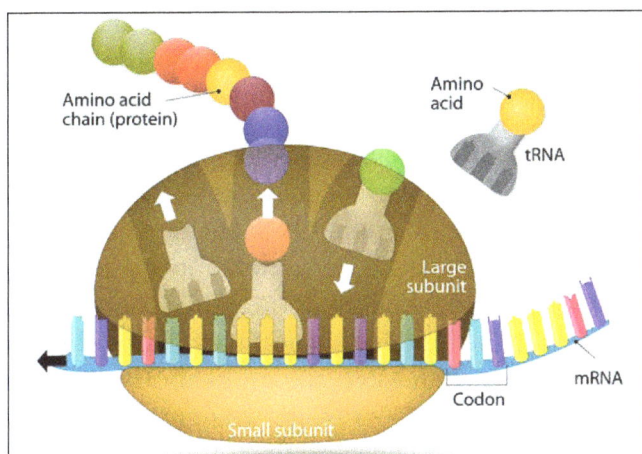

A ribosome is a biological machine.

A roadmap for the development of MNT is an objective of a broadly based technology project led by Battelle (the manager of several U.S. National Laboratories) and the Foresight Institute. The roadmap was originally scheduled for completion by late 2006, but was released in January 2008. The Nanofactory Collaboration is a more focused ongoing effort involving 23 researchers from 10 organizations and 4 countries that is developing a practical research agenda specifically aimed at positionally-controlled diamond mechanosynthesis and diamondoid nanofactory development. In August 2005, a task force consisting of 50+ international experts from various fields was organized by the Center for Responsible Nanotechnology to study the societal implications of molecular nanotechnology.

Projected Applications and Capabilities

Smart Materials and Nanosensors

One proposed application of MNT is so-called smart materials. This term refers to any sort of material designed and engineered at the nanometer scale for a specific task. It encompasses a wide variety of possible commercial applications. One example would be materials designed to respond differently to various molecules; such a capability could lead, for example, to artificial drugs which would recognize and render inert specific viruses. Another is the idea of self-healing structures, which would repair small tears in a surface naturally in the same way as self-sealing tires or human skin.

A MNT nanosensor would resemble a smart material, involving a small component within a larger machine that would react to its environment and change in some fundamental, intentional way. A very simple example: a photosensor might passively measure the incident light and discharge its absorbed energy as electricity when the light passes above or below a specified threshold, sending a signal to a larger machine. Such a sensor would supposedly cost less and use less power than a conventional sensor, and yet function usefully in all the same applications — for example, turning on parking lot lights when it gets dark. While smart materials and nanosensors both exemplify useful applications of MNT, they pale in comparison with the complexity of the technology most popularly associated with the term: the replicating nanorobot.

Replicating Nanorobots

MNT nanofacturing is popularly linked with the idea of swarms of coordinated nanoscale robots working together, a popularization of an early proposal by K. Eric Drexler in his 1986 discussions of MNT, but superseded in 1992. In this early proposal, sufficiently capable nanorobots would construct more nanorobots in an artificial environment containing special molecular building blocks.

Critics have doubted both the feasibility of self-replicating nanorobots and the feasibility of control if self-replicating nanorobots could be achieved: they cite the possibility of mutations removing any control and favoring reproduction of mutant pathogenic variations. Advocates address the first doubt by pointing out that the first macroscale autonomous machine replicator, made of Lego blocks, was built and operated experimentally in 2002. While there are sensory advantages present at the macroscale compared to the limited sensorium available at the nanoscale, proposals for positionally controlled nanoscale mechanosynthetic fabrication systems employ dead reckoning of tooltips combined with reliable reaction sequence design to ensure reliable results, hence a limited sensorium is no handicap; similar considerations apply to the positional assembly of small nanoparts. Advocates address the second doubt by arguing that bacteria are (of necessity) evolved to evolve, while nanorobot mutation could be actively prevented by common error-correcting techniques. Similar ideas are advocated in the Foresight Guidelines on Molecular Nanotechnology, and a map of the 137-dimensional replicator design space recently published by Freitas and Merkle provides numerous proposed methods by which replicators could, in principle, be safely controlled by good design.

However, the concept of suppressing mutation raises the question: How can design evolution occur at the nanoscale without a process of random mutation and deterministic selection? Critics argue that MNT advocates have not provided a substitute for such a process of evolution in this nanoscale arena where conventional sensory-based selection processes are lacking. The limits of the sensorium available at the nanoscale could make it difficult or impossible to winnow successes from failures. Advocates argue that design evolution should occur deterministically and strictly under human control, using the conventional engineering paradigm of modeling, design, prototyping, testing, analysis, and redesign.

In any event, since 1992 technical proposals for MNT do not include self-replicating nanorobots, and recent ethical guidelines put forth by MNT advocates prohibit unconstrained self-replication.

Medical Nanorobots

One of the most important applications of MNT would be medical nanorobotics or nanomedicine, an area pioneered by Robert Freitas in numerous books and papers. The ability to design, build, and deploy large numbers of medical nanorobots would, at a minimum, make possible the rapid elimination of disease and the reliable and relatively painless recovery from physical trauma. Medical nanorobots might also make possible the convenient correction of genetic defects, and help to ensure a greatly expanded lifespan. More controversially, medical nanorobots might be used to augment natural human capabilities. One study has reported on the conditions like tumors, arteriosclerosis, blood clots leading to stroke, accumulation of scar tissue and localized pockets of infection can be possibly be addressed by employing medical nanorobots.

Utility Fog

Another proposed application of molecular nanotechnology is "utility fog" — in which a cloud of networked microscopic robots (simpler than assemblers) would change its shape and properties to form macroscopic objects and tools in accordance with software commands. Rather than modify the current practices of consuming material goods in different forms, utility fog would simply replace many physical objects.

Diagram of a 100 micrometer foglet.

Phased-array Optics

Yet another proposed application of MNT would be phased-array optics (PAO). However, this appears to be a problem addressable by ordinary nanoscale technology. PAO would use the principle of phased-array millimeter technology but at optical wavelengths. This would permit the duplication of any sort of optical effect but virtually. Users could request holograms, sunrises and sunsets, or floating lasers as the mood strikes.

Potential Social Impacts

Molecular manufacturing is a potential future subfield of nanotechnology that would make it possible to build complex structures at atomic precision. Molecular manufacturing requires significant advances in nanotechnology, but once achieved could produce highly advanced products at low costs and in large quantities in nanofactories weighing a kilogram or more. When nanofactories gain the ability to produce other nanofactories production may only be limited by relatively abundant factors such as input materials, energy and software.

The products of molecular manufacturing could range from cheaper, mass-produced versions of known high-tech products to novel products with added capabilities in many areas of application. Some applications that have been suggested are advanced smart materials, nanosensors, medical nanorobots and space travel. Additionally, molecular manufacturing could be used to cheaply produce highly advanced, durable weapons, which is an area of special concern regarding the impact of nanotechnology. Being equipped with compact computers and motors these could be increasingly autonomous and have a large range of capabilities.

According to Chris Phoenix and Mike Treder from the Center for Responsible Nanotechnology as well as Anders Sandberg from the Future of Humanity Institute molecular manufacturing is the application of nanotechnology that poses the most significant global catastrophic risk. Several nanotechnology researchers state that the bulk of risk from nanotechnology comes from the potential to lead to war, arms races and destructive global government. Several reasons have been suggested why the availability of nanotech weaponry may with significant likelihood lead to unstable arms races (compared to e.g. nuclear arms races): (1) A large number of players may be tempted to enter the race since the threshold for doing so is low; (2) the ability to make weapons with molecular manufacturing will be cheap and easy to hide; (3) therefore lack of insight into the other parties' capabilities can tempt players to arm out of caution or to launch preemptive strikes; (4) molecular manufacturing may reduce dependency on international trade, a potential peace-promoting factor; (5) wars of aggression may pose a smaller economic threat to the aggressor since manufacturing is cheap and humans may not be needed on the battlefield.

Since self-regulation by all state and non-state actors seems hard to achieve, measures to mitigate war-related risks have mainly been proposed in the area of international cooperation. International infrastructure may be expanded giving more sovereignty to the international level. This could help coordinate efforts for arms control. International institutions dedicated specifically to nanotechnology (perhaps analogously to the International Atomic Energy Agency IAEA) or general arms control may also be designed. One may also jointly make differential technological progress on defensive technologies, a policy that players should usually favour. The Center for Responsible Nanotechnology also suggest some technical restrictions. Improved transparency regarding technological capabilities may be another important facilitator for arms-control.

A grey goo is another catastrophic scenario, which was proposed by Eric Drexler has been analyzed by Freitas in "Some Limits to Global Ecophagy by Biovorous Nanoreplicators, with Public Policy Recommendations" and has been a theme in mainstream media and fiction. This scenario involves tiny self-replicating robots that consume the entire biosphere using it as a source of energy and building blocks. Nanotech experts including Drexler now discredit the scenario. According to Chris Phoenix a "So-called grey goo could only be the product of a deliberate and difficult engineering process, not an accident". With the advent of nano-biotech, a different scenario called green goo has been forwarded. Here, the malignant substance is not nanobots but rather self-replicating biological organisms engineered through nanotechnology.

Benefits of Nanotechnology

Nanotechnology or molecular nanotechnology will let us continue the historical trends in manufacturing right up to the fundamental limits imposed by physical law. It will let us make remarkably powerful molecular computers. It will let us make materials over fifty times lighter than steel or aluminium alloy but with the same strength. We'll be able to make jets, rockets, cars or even chairs that, by today's standards, would be remarkably light, strong, and inexpensive. Molecular surgical tools, guided by molecular computers and injected into the blood stream could find and destroy cancer cells or invading bacteria, unclog arteries, or provide oxygen when the circulation is impaired.

Nanotechnology will replace our entire manufacturing base with a new, radically more precise, radically less expensive, and radically more flexible way of making products. The aim is not simply to replace today's computer chip making plants, but also to replace the assembly lines for cars, televisions, telephones, books, surgical tools, missiles, bookcases, airplanes, tractors, and all the rest. The objective is a pervasive change in manufacturing, a change that will leave virtually no product untouched. Economic progress and military readiness in the 21st Century will depend fundamentally on maintaining a competitive position in nanotechnology.

Despite the current early developmental status of nanotechnology and molecular nanotechnology, much concern surrounds MNT's anticipated impact on economics and on law. Whatever the exact effects, MNT, if achieved, would tend to reduce the scarcity of manufactured goods and make many more goods (such as food and health aids) manufacturable.

MNT should make possible nanomedical capabilities able to cure any medical condition not already cured by advances in other areas. Good health would be common, and poor health of any form would be as rare as smallpox and scurvy are today. Even cryonics would be feasible, as cryopreserved tissue could be fully repaired.

Risks Realated to Nanotechnology

Molecular nanotechnology is one of the technologies that some analysts believe could lead to a technological singularity. Some feel that molecular nanotechnology would have daunting risks. It conceivably could enable cheaper and more destructive conventional weapons. Also, molecular nanotechnology might permit weapons of mass destruction that could self-replicate, as viruses and cancer cells do when attacking the human body. Commentators generally agree that, in the event molecular nanotechnology were developed, its self-replication should be permitted only under very controlled or "inherently safe" conditions.

A fear exists that nanomechanical robots, if achieved, and if designed to self-replicate using naturally occurring materials (a difficult task), could consume the entire planet in their hunger for raw materials, or simply crowd out natural life, out-competing it for energy (as happened historically when blue-green algae appeared and outcompeted earlier life forms). Some commentators have referred to this situation as the "grey goo" or "ecophagy" scenario. K. Eric Drexler considers an accidental "grey goo" scenario extremely unlikely and says so in later editions of *Engines of Creation*.

In light of this perception of potential danger, the Foresight Institute, founded by Drexler, has prepared a set of guidelines for the ethical development of nanotechnology. These include the banning of free-foraging self-replicating pseudo-organisms on the Earth's surface, at least, and possibly in other places.

NANOROBOTICS

Nanorobotics is the technology of creating machines or robots close to the microscopic scale of a nanometer (10–9 meters). Nanorobotics refers to nanotechnology – an engineering discipline for designing and building nanorobots. These devices range from 0.1-10 micrometers and are made up of nano scale or molecular components. As no artificial, non-biological Nano robots have yet been created, they remain a pretending concept. The names nanorobots, nanoids, nanites or nanomites have also been used to describe these hypothetical devices.

Nano robots can be used in different application areas such as medicine and space technology. Nowadays, these nanorobots play a crucial role in the field of Bio-Medicine, particularly for the treatment of cancer, cerebral Aneurysm, removal of kidney stones, elimination of defected parts in the DNA structure, and for some other treatments that need utmost support to save human lives.

Nanorobots.

Nanorobots are nano devices used for the purpose of maintaining and protecting the human body against pathogens. Nanorobots are implemented by using several components such as sensors, actuators, control, power, communication and by interfacing cross-special scales between organic inorganic systems.

The development of nanorobots is done by using various approaches such as:

Biochip

The combination of nanotechnology, photo-lithography and new biomaterials, can be considered as a possible way required for designing technology to develop nanorobots for medical applications such as diagnosis and drug delivery. This realistic approach in designing nanorobots is a methodology which is used in the electronic industries.

Nubots

Nubot is an acronym for "nucleic acid robots." Nubots are manmade robotics devices at the Nanoscale. Representative nubots includes numerous Deoxy Nucleic Acid walkers reported by Ned Seeman's group at NYU, Niles Pierce's group at Caltech, John Reif's group at Duke University, Chengde Mao's group at Purdue, and Andrew Turberfield's group at the University of Oxford.

Positional Nanoassembly

In the year 2000, Robert Frietas and Ralph Merkle found nanofactory collaboration which is an ongoing effort consisting of ten organizations with 23 researchers from four countries. This collaboration aims at developing positionally controlled mechanosynthesis and diamondoid nanofactory which is capable of constructing a diamondoid medical nanorobot.

Usage of Bacteria

This approach makes use of biological microorganisms, such as Escherichia coil bacteria. So this model uses a flagellum for propulsion purpose. The use of electromagnetic fields is to control the motion of biological integrated device and its limited applications.

Nanorobots Applications

Nanorobotics in Surgery

Surgical nanorobots are introduced into the human body through vascular systems and other cavities. Surgical nanorobots act as semi-autonomous on-site surgeon inside the human body and are programmed or directed by a human surgeon. This programmed surgical nanorobot performs various functions like searching for pathogens, and then diagnosis and correction of lesions by nano-manipulation synchronized by an on-board computer while conserving and contacting with the supervisory surgeon through coded ultrasound signals.

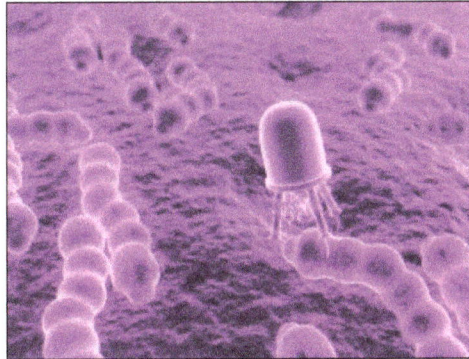
Nanorobotics in Surgery.

Nowadays, the earlier forms of cellular nano-surgery are being explored. For example, a micropipette rapidly vibrating at a frequency of 100 Hz micropipette comparatively less than 1 micron tip diameter is used to cut dendrites from single neurons. This process is not ought to damage the cell capability.

Diagnosis and Testing

Medical nanorobots are used for the purpose of diagnosis, testing and monitoring of microorganisms, tissues and cells in the blood stream. These nanorobots are capable of noting down the record, and report some vital signs such as temperature, pressure and immune system's parameters of different parts of the human body continuously.

Nanorobotics in Gene Therapy

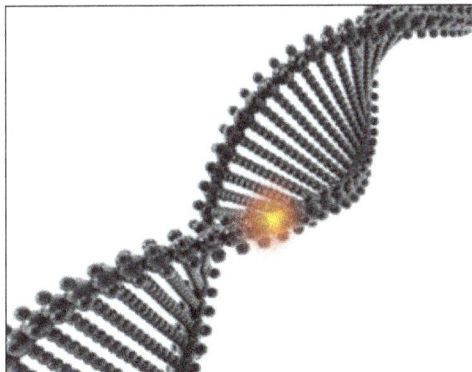
Nanorobotics in Gene Therapy.

Nanorobots are also applicable in treating genetic diseases, by relating the molecular structures of DNA and proteins in the cell. The modifications and irregularities in the DNA and protein sequences are then corrected (edited). The chromosomal replacement therapy is very efficient compared to the cell repair. An assembled repair vessel is inbuilt in the human body to perform the maintenance of genetics by floating inside the nucleus of a cell.

Supercoil of DNA when enlarged within its lower pair of robotic arms, the nanomachine pulls the strand which is unwounded for analysis; meanwhile the upper arms detach the proteins from the chain. The information which is stored in the large nanocomputer's database is placed outside the nucleus and compared with the molecular structures of both DNA and proteins that are connected through communication link to cell repair ship. Abnormalities found in the structures are corrected, and the proteins reattached to the Deoxy Nucleic Acid chain once again reforms into their original form.

Nanorobots in Cancer Detection and Treatment

The current stages of medical technologies and therapy tools are used for the successful treatment of cancer. The important aspect to achieve a successful treatment is based on the improvement of efficient drug delivery to decrease the side-effects from the chemotherapy.

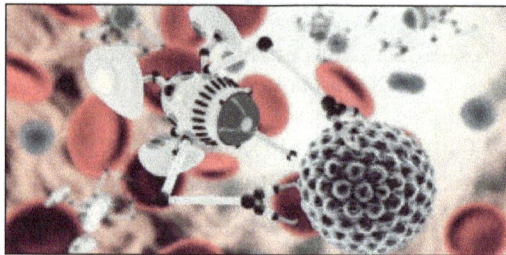

Nanorobots in Cancer Detection and Treatment.

Nanorobots with embedded chemical biosensors are used for detecting the tumor cells in early stages of cancer development inside a patient's body. Nanosensors are also utilized to find the intensity of E-cadherin signals.

Nanodentistry is one of the topmost applications as nanorobots help in different processes involved in dentistry.These nanorobots are helpful in desensitizing tooth, oral anesthesia, straightening of irregular set of teeth and improvement of the teeth durability, major tooth repairs and improvement of appearance of teeth, etc.

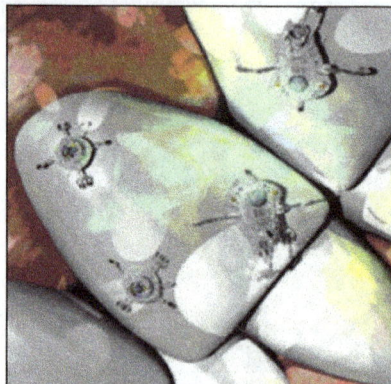

Nanodentistry.

Nanorobots can also be used as ancillary devices for processing different chemical reactions in the affected organs. These robots are also useful for monitoring and controlling the glucose levels in diabetic patients.

GREEN NANOTECHNOLOGY

Green nanotechnology refers to the use of nanotechnology to enhance the environmental sustainability of processes producing negative externalities. It also refers to the use of the products of nanotechnology to enhance sustainability. It includes making green nano-products and using nano-products in support of sustainability.

Green nanotechnology has been described as the development of clean technologies, "to minimize potential environmental and human health risks associated with the manufacture and use of nanotechnology products, and to encourage replacement of existing products with new nano-products that are more environmentally friendly throughout their lifecycle."

Green nanotechnology has two goals: producing nanomaterials and products without harming the environment or human health, and producing nano-products that provide solutions to environmental problems. It uses existing principles of green chemistry and green engineering to make nanomaterials and nano-products without toxic ingredients, at low temperatures using less energy and renewable inputs wherever possible, and using lifecycle thinking in all design and engineering stages.

In addition to making nanomaterials and products with less impact to the environment, green nanotechnology also means using nanotechnology to make current manufacturing processes for non-nano materials and products more environmentally friendly. For example, nanoscale membranes can help separate desired chemical reaction products from waste materials. Nanoscale catalysts can make chemical reactions more efficient and less wasteful. Sensors at the nanoscale can form a part of process control systems, working with nano-enabled information systems. Using alternative energy systems, made possible by nanotechnology, is another way to "green" manufacturing processes.

The second goal of green nanotechnology involves developing products that benefit the environment either directly or indirectly. Nanomaterials or products directly can clean hazardous waste sites, desalinate water, treat pollutants, or sense and monitor environmental pollutants. Indirectly, lightweight nanocomposites for automobiles and other means of transportation could save fuel and reduce materials used for production; nanotechnology-enabled fuel cells and light-emitting diodes (LEDs) could reduce pollution from energy generation and help conserve fossil fuels; self-cleaning nanoscale surface coatings could reduce or eliminate many cleaning chemicals used in regular maintenance routines; and enhanced battery life could lead to less material use and less waste. Green Nanotechnology takes a broad systems view of nanomaterials and products, ensuring that unforeseen consequences are minimized and that impacts are anticipated throughout the full life cycle.

Applications

Solar Cells

Research is underway to use nanomaterials for purposes including more efficient solar cells, practical fuel cells, and environmentally friendly batteries. The most advanced nanotechnology projects related to energy are: storage, conversion, manufacturing improvements by reducing materials and process rates, energy saving (by better thermal insulation for example), and enhanced renewable energy sources.

One major project that is being worked on is the development of nanotechnology in solar cells. Solar cells are more efficient as they get tinier and solar energy is a renewable resource. The price per watt of solar energy is lower than one dollar.

Research is ongoing to use nanowires and other nanostructured materials with the hope of to create cheaper and more efficient solar cells than are possible with conventional planar silicon solar cells. Another example is the use of fuel cells powered by hydrogen, potentially using a catalyst consisting of carbon supported noble metal particles with diameters of 1–5 nm. Materials with small nanosized pores may be suitable for hydrogen storage. Nanotechnology may also find applications in batteries, where the use of nanomaterials may enable batteries with higher energy content or supercapacitors with a higher rate of recharging.

Nanotechnology is already used to provide improved performance coatings for photovoltaic (PV) and solar thermal panels. Hydrophobic and self-cleaning properties combine to create more efficient solar panels, especially during inclement weather. PV covered with nanotechnology coatings are said to stay cleaner for longer to ensure maximum energy efficiency is maintained.

Nanoremediation and Water Treatment

Nanotechnology offers the potential of novel nanomaterials for the treatment of surface water, groundwater, wastewater, and other environmental materials contaminated by toxic metal ions, organic and inorganic solutes, and microorganisms. Due to their unique activity toward recalcitrant contaminants, many nanomaterials are under active research and development for use in the treatment of water and contaminated sites.

The present market of nanotech-based technologies applied in water treatment consists of reverse osmosis(RO), nanofiltration, ultrafiltration membranes. Indeed, among emerging products one can name nanofiber filters, carbon nanotubes and various nanoparticles. Nanotechnology is expected to deal more efficiently with contaminants which convectional water treatment systems struggle to treat, including bacteria, viruses and heavy metals. This efficiency generally stems from the very high specific surface area of nanomaterials which increases dissolution, reactivity and sorption of contaminants.

Environmental Remediation

Nanoremediation is the use of nanoparticles for environmental remediation. Nanoremediation has been most widely used for groundwater treatment, with additional extensive research in wastewater treatment. Nanoremediation has also been tested for soil and sediment cleanup. Even more preliminary research is exploring the use of nanoparticles to remove toxic materials from gases.

Some nanoremediation methods, particularly the use of nano zerovalent iron for groundwater cleanup, have been deployed at full-scale cleanup sites. Nanoremediation is an emerging industry;

by 2009, nanoremediation technologies had been documented in at least 44 cleanup sites around the world, predominantly in the United States. During nanoremediation, a nanoparticle agent must be brought into contact with the target contaminant under conditions that allow a detoxifying or immobilizing reaction. This process typically involves a pump-and-treat process or *in situ* application. Other methods remain in research phases.

Scientists have been researching the capabilities of buckminsterfullerene in controlling pollution, as it may be able to control certain chemical reactions. Buckminsterfullerene has been demonstrated as having the ability of inducing the protection of reactive oxygen species and causing lipid peroxidation. This material may allow for hydrogen fuel to be more accessible to consumers.

Water Cleaning Technology

In 2017 the RingwooditE Co Ltd was formed in order to explore Thermonuclear Trap Technology (TTT) for the purpose of cleaning all sources of water from pollution and toxic contents. This patented nanotechnology uses a high pressure and temperature chamber to separate isotopes that should by nature not be in drinking water to pure drinking water, as to the by the WHO´s established classification. This method has been developed by among others, by professor Vladimir Afanasiew, at the Moscow Nuclear Institution. This technology is targeted to clean Sea, river, lake and landfill waste waters. It even removes radioactive isotopes from the sea water, after Nuclear Power Stations catastrophes and cooling water plant towers. By this technology pharmaca rests are being removed as well as narcotics and tranquilizers. Bottom layers and sides at lake and rivers can be returned, after being cleaned. Machinery used for this purpose are much similar to those of deep sea mining. Removed waste items are being sorted by the process, and can be re used as raw material for other industrial production.

Water Filtration

Nanofiltration is a relatively recent membrane filtration process used most often with low total dissolved solids water such as surface water and fresh groundwater, with the purpose of softening (polyvalent cation removal) and removal of disinfection by-product precursors such as natural organic matter and synthetic organic matter. Nanofiltration is also becoming more widely used in food processing applications such as dairy, for simultaneous concentration and partial (monovalent ion) demineralisation.

Nanofiltration is a membrane filtration based method that uses nanometer sized cylindrical through-pores that pass through the membrane at a 90°. Nanofiltration membranes have pore sizes from 1-10 Angstrom, smaller than that used in microfiltration and ultrafiltration, but just larger than that in reverse osmosis. Membranes used are predominantly created from polymer thin films. Materials that are commonly used include polyethylene terephthalate or metals such as aluminum. Pore dimensions are controlled by pH, temperature and time during development with pore densities ranging from 1 to 106 pores per cm². Membranes made from polyethylene terephthalate and other similar materials, are referred to as "track-etch" membranes, named after the way the pores on the membranes are made. "Tracking" involves bombarding the polymer thin film with high energy particles. This results in making tracks that are chemically developed into the membrane, or "etched" into the membrane, which are the pores. Membranes created from metal such as alumina membranes, are made by electrochemically growing a thin layer of aluminum oxide from aluminum metal in an acidic medium.

Some water-treatment devices incorporating nanotechnology are already on the market, with more in development. Low-cost nanostructured separation membranes methods have been shown to be effective in producing potable water in a recent study.

Nanotech to Disinfect Water

Nanotechnology provides an alternative solution to clean germs in water, a problem that has been getting worse due to the population explosion, growing need for clean water and the emergence of additional pollutants. One of the alternatives offered is antimicrobial nanotechnology stated that several nanomaterials showed strong antimicrobial properties through diverse mechanisms, such as photocatalytic production of reactive oxygen species that damage cell components and viruses. There is also the case of the synthetically-fabricated nanometallic particles that produce antimicrobial action called oligodynamic disinfection, which can inactivate microorganisms at low concentrations. Commercial purification systems based on titanium oxide photocatalysis also currently exist and studies show that this technology can achieve complete inactivation of fecal coliforms in 15 minutes once activated by sunlight.

There are four classes of nanomaterials that are employed for water treatment and these are dendrimers, zeolites, carbonaceous nanomaterials, and metals containing nanoparticles. The benefits of the reduction of the size of the metals (e.g. silver, copper, titanium, and cobalt) to the nanoscale such as contact efficiency, greater surface area, and better elution properties.

Cleaning up Oil Spills

The U.S. Environmental Protection Agency (EPA) documents more than ten thousand oil spills per year. Conventionally, biological, dispersing, and gelling agents are deployed to remedy oil spills. Although, these methods have been used for decades, none of these techniques can retrieve the irreplaceable lost oil. However, nanowires can not only swiftly clean up oil spills but also recover as much oil as possible. These nanowires form a mesh that absorbs up to twenty times its weight in hydrophobic liquids while rejecting water with its water repelling coating. Since the potassium manganese oxide is very stable even at high temperatures, the oil can be boiled off the nanowires and both the oil and the nanowires can then be reused.

In 2005, Hurricane Katrina damaged or destroyed more than thirty oil platforms and nine refineries. The Interface Science Corporation successfully launched a new oil remediation and recovery application, which used the water repelling nanowires to clean up the oil spilled by the damaged oil platforms and refineries.

Removing Plastics from Oceans

One innovation of green nanotechnology that is currently under development are nanomachines modeled after a bacteria bioengineered to consume plastics, *Ideonella sakaiensis*. These nano-machines are able to decompose plastics dozens of times faster than the bioengineered bacteria not only because of their increased surface area but also because of the fact that the energy released from decomposing the plastic is used to fuel the nano-machines.

Air Pollution Control

In addition to water treatment and environmental remediation, nanotechnology is currently improving air quality. Nanoparticles can be engineered to catalyze, or hasten, the reaction to transform environmentally pernicious gases into harmless ones. For example, many industrial factories that produce large amounts harmful gases employ a type of nanofiber catalyst made of magnesium oxide (Mg_2O) to purify dangerous organic substances in the smoke. Although chemical catalysts already exist in the gaseous vapors from cars, nanotechnology has a greater chance of reacting with the harmful substances in the vapors. This greater probability comes from the fact that nanotechnology can interact with more particles because of its greater surface area.

Nanotechnology has been used to remediate air pollution including car exhaust pollution, and potentially greenhouse gases due to its high surface area. Based on research done by the Environmental Science Pollution Research International, nanotechnology can specifically help to treat carbon-based nanoparticles, greenhouse gases, and volatile organic compounds. There is also work being done to develop antibacterial nanoparticles, metal oxide nanoparticles, and amendment agents for phytoremediation processes. Nanotechnology can also give the possibility of preventing air pollution in the first place due to its extremely small scale. Nanotechnology has been accepted as a tool for many industrial and domestic fields like gas monitoring systems, fire and toxic gas detectors, ventilation control, breath alcohol detectors and many more. Other sources state that nanotechnology has the potential to develop the pollutants sensing and detection methods that already exist. The ability to detect pollutants and sense unwanted materials will be heightened by the large surface area of nanomaterials and their high surface energy. The World Health Organization declared in 2014 that air contamination caused around 7 million deaths in 2012. This new technology could be an essential asset to this epidemic. The three ways that nanotechnology is being used to treat air pollution are nano-adsoptive materials, degradation by nanocatalysis, and filtration/separation by nanofilters. Nanoscale adsorbents being the main alleviator for many air pollution difficulties. Their structure permits a great interaction with organic compounds as well as increased selectivity and stability in maximum adsorption capacity. Other advantages include high electrical and thermal conductivities, high strength, high hardness. Target pollutants that can be targeted by nanomolecules are NO_x, CO_2, NH_3, N_2, VOCs, Isopropyl vapor, CH_3OH gases, N_2O, H_2S. Carbon nanotubes specifically remove particles in many ways. One method is by passing them through the nanotubes where the molecules are oxidized; the molecules then are adsorbed on a nitrate species. Carbon nanotubes with amine groups provide numerous chemical sites for carbon dioxide adsorption at low temperature ranges of 20°-100° degrees Celsius. Van der waals forces and π-π interactions also are used to pull molecules onto surface functional groups. Fullerene can be used to rid of carbon dioxide pollution due to its high adsorption capacity. Graphene nanotubes have functional groups that adsorb gases. There are plenty of nanocatalysts that can be used for air pollution reduction and air quality. Some of these materials include TiO_2, Vanadium, Platinum, Palladium, Rhodium, and Silver. Catalytic industrial emission reduction, car exhaust reduction, and air purification are just some of the major thrusts that these nanomaterials are being utilized within. Certain applications are not widely spread, but other are more popular. Indoor air pollution is barely on the market yet, but it is being developed more efficiently due to complications with health effects. Car exhaust emission reduction is widely used in diesel fueled automobiles currently being one of the more popular applications. Industrial emission reduction is also widely used. It is n integral method specifically at coal fired power plants as well as refineries. According to Macrothink Institute process control, ecosystem monitoring is becoming more sensitive and precise due to nanotechnology development and research. These methods are analyzed and reviewed using SEM imaging to ensure its usefulness and accuracy.

Additionally, research is currently being conducted to find out if nanoparticles can be engineered to separate car exhaust from methane or carbon dioxide, which has been known to damage the Earth's ozone layer. In fact, John Zhu, a professor at the University of Queensland, is exploring the creation of a carbon nanotube(CNT) which can trap greenhouse gases hundreds of times more efficiently than current methods can.

Nanotechnology for Sensors

Perpetual exposure to heavy metal pollution and particulate matter will lead to health concerns such as lung cancer, heart conditions, and even motor neuron diseases. However, humanity's ability to shield themselves from these health problems can be improved by accurate and swift nano-contact-sensors able to detect pollutants at the atomic level. These nanocontact sensors do not require much energy to detect metal ions or radioactive elements. Additionally, they can be made in automatic mode so that they can be readably used at any given moment. Additionally, these nanocontact sensors are energy and cost effective since they are composed with conventional microelectronic manufacturing equipment using electrochemical techniques.

Some examples of nano-based monitoring include:

1. Functionalized nanoparticles able to form anionic oxidants bonding thereby allowing the detection of carcinogenic substances at very low concentrations.

2. Polymer nanospheres have been developed to measure organic contaminates in very low concentrations.

3. "Peptide nanoelectrodes have been employed based on the concept of thermocouple. In a nano-distance separation gap, a peptide molecule is placed to form a molecular junction. When a specific metal ion is bound to the gap; the electrical current will result conductance in a unique value. Hence the metal ion will be easily detected."

4. Composite electrodes, a mixture of nanotubes and copper, have been created to detect substances such as organophosphorus pesticides, carbohydrates and other woods pathogenic substances in low concentrations.

Concerns

Although green nanotechnology poses many advantages over traditional methods, there is still much debate about the concerns brought about by nanotechnology. For example, since the nanoparticles are small enough to be absorbed into skin and inhaled, countries are mandating that additional research revolving around the impact of nanotechnology on organisms be heavily studied. In fact, the field of eco-nanotoxicology was founded solely to study the effect of nanotechnology on earth and all of its organisms. At the moment, scientists are unsure of what will happen when nanoparticles seep into soil and water, but organizations, such as NanoImpactNet, have set out to study these effects.

NANOREMEDIATION

Nanoremediation is the use of nanoparticles for environmental remediation. It is being explored to treat ground water, wastewater, soil, sediment, or other contaminated environmental materials. Nanoremediation is an emerging industry; by 2009, nanoremediation technologies had been documented in at least 44 cleanup sites around the world, predominantly in the United States. In Europe, nanoremediation is being investigated by the EC funded NanoRem Project. A report produced by the NanoRem consortium has identified around 70 nanoremediation projects worldwide at pilot or full scale. During nanoremediation, a nanoparticle agent must be brought into contact with the target contaminant under conditions that allow a detoxifying or immobilizing reaction. This process typically involves a pump-and-treat process or *in situ* application.

Some nanoremediation methods, particularly the use of nano zero-valent iron for groundwater cleanup, have been deployed at full-scale cleanup sites. Other methods remain in research phases.

Applications

Nanoremediation has been most widely used for groundwater treatment, with additional extensive research in wastewater treatment. Nanoremediation has also been tested for soil and sediment cleanup. Even more preliminary research is exploring the use of nanoparticles to remove toxic materials from gases.

Groundwater Remediation

Currently, groundwater remediation is the most common commercial application of nanoremediation technologies. Using nanomaterials, especially zero-valent metals (ZVMs), for groundwater remediation is an emerging approach that is promising due to the availability and effectiveness of many nanomaterials for degrading or sequestering contaminants.

Nanotechnology offers the potential to effectively treat contaminants *in situ*, avoiding excavation or the need to pump contaminated water out of the ground. The process begins with nanoparticles being injected into a contaminated aquifer via an injection well. The nanoparticles are then transported by groundwater flow to the source of contamination. Upon contact, nanoparticles can sequester contaminants (via adsorption or complexation), immobilizing them, or they can degrade the contaminants to less harmful compounds. Contaminant transformations are typically redox reactions. When the nanoparticle is the oxidant or reductant, it is considered reactive.

The ability to inject nanoparticles to the subsurface and transport them to the contaminant source is imperative for successful treatment. Reactive nanoparticles can be injected into a well where they will then be transported down gradient to the contaminated area. Drilling and packing a well is quite expensive. Direct push wells cost less than drilled wells and are the most often used delivery tool for remediation with nanoiron. A nanoparticle slurry can be injected along the vertical range of the probe to provide treatment to specific aquifer regions.

Surface Water Treatment

The use of various nanomaterials, including carbon nanotubes and TiO_2, shows promise for treatment of surface water, including for purification, disinfection, and desalination. Target contaminants in surface waters include heavy metals, organic contaminants, and pathogens. In this context, nanoparticles may be used as sorbents, as reactive agents (photocatalysts or redox agents), or in membranes used for nanofiltration.

Trace Contaminant Detection

Nanoparticles may assist in detecting trace levels of contaminants in field settings, contributing to effective remediation. Instruments that can operate outside of a laboratory often are not sensitive enough to detect trace contaminants. Rapid, portable, and cost-effective measurement systems for trace contaminants in groundwater and other environmental media would thus enhance contaminant detection and cleanup. One potential method is to separate the analyte from the sample and concentrate them to a smaller volume, easing detection and measurement. When small quantities of solid sorbents are used to absorb the target for concentration, this method is referred to as solid-phase microextraction.

With their high reactivity and large surface area, nanoparticles may be effective sorbents to help concentrate target contaminants for solid-phase microextraction, particularly in the form of self-assembled monolayers on mesoporous supports. The mesoporous silica structure, made through a surfactant templated sol-gel process, gives these self-assembled monolayers high surface area and a rigid open pore structure. This material may be an effective sorbent for many targets, including heavy metals such as mercury, lead, and cadmium, chromate and arsenate, and radionuclides such as ^{99}Tc, ^{137}CS, uranium, and the actinides.

Mechanism

The small size of nanoparticles leads to several characteristics that may enhance remediation. Nanomaterials are highly reactive because of their high surface area per unit mass. Their small particle size also allows nanoparticles to enter small pores in soil or sediment that larger particles might not penetrate, granting them access to contaminants sorbed to soil and increasing the likelihood of contact with the target contaminant.

Because nanomaterials are so tiny, their movement is largely governed by Brownian motion as compared to gravity. Thus, the flow of groundwater can be sufficient to transport the particles. Nanoparticles then can remain suspended in solution longer to establish an *in situ* treatment zone.

Once a nanoparticle contacts the contaminant, it may degrade the contaminant, typically through a redox reaction, or adsorb to the contaminant to immobilize it. In some cases, such as with magnetic nano-iron, adsorbed complexes may be separated from the treated substrate, removing the contaminant. Target contaminants include organic molecules such as pesticides or organic solvents and metals such as arsenic or lead. Some research is also exploring the use of nanoparticles to remove excessive nutrients such as nitrogen and phosphorus.

Materials

A variety of compounds, including some that are used as macro-sized particles for remediation, are being studied for use in nanoremediation. These materials include zero-valent metals like zero-valent iron, calcium carbonate, carbon-based compounds such as graphene or carbon nanotubes, and metal oxides such as titanium dioxide and iron oxide.

Nano Zero-valent Iron

As of 2012, nano zero-valent iron (nZVI) was the nanoscale material most commonly used in bench and field remediation tests. nZVI may be mixed or coated with another metal, such as palladium, silver, or copper, that acts as a catalyst in what is called a bimetallic nanoparticle. nZVI may also be emulsified with a surfactant and an oil, creating a membrane that enhances the nanoparticle's ability to interact with hydrophobic liquids and protects it against reactions with materials dissolved in water. Commercial nZVI particle sizes may sometimes exceed true "nano" dimensions (100 nm or less in diameter).

nZVI appears to be useful for degrading organic contaminants, including chlorinated organic compounds such as polychlorinated biphenyls (PCBs) and trichloroethene (TCE), as well as immobilizing or removing metals. nZVI and other nanoparticles that do not require light can be injected belowground into the contaminated zone for *in situ* groundwater remediation and, potentially, soil remediation.

nZVI nanoparticles can be prepared by using sodium borohydride as the key reductant. $NaBH_4$ (0.2 M) is added into $FeCl_3 \cdot 6H_2$ (0.05 M) solution (~1:1 volume ratio). Ferric iron is reduced via the following reaction:

$$4Fe^{3+} + 3BH_4^- + 9H_2O \rightarrow 4Fe^0 + 3H_2BO_3^- + 12H^+ + 6H_2$$

Palladized Fe particles are prepared by soaking the nanoscale iron particles with an ethanol solution of 1wt% of palladium acetate ($[Pd(C_2H_3O_2)2]_3$). This causes the reduction and deposition of Pd on the Fe surface:

$$Pd^{2+} + Fe^0 \rightarrow Pd^0 + Fe^{2+}$$

Similar methods may be used to prepared Fe/Pt, Fe/Ag, Fe/Ni, Fe/Co, and Fe/Cu bimetallic particles. With the above methods, nanoparticles of diameter 50-70 nm may be produced. The average specific surface area of Pd/Fe particles is about 35 m²/g. Ferrous iron salt has also been successfully used as the precursor.

Titanium Dioxide

Titanium dioxide (TiO_2) is also a leading candidate for nanoremediation and wastewater treatment, although as of 2010 it is reported to have not yet been expanded to full-scale commercialization. When exposed to ultraviolet light, such as in sunlight, titanium dioxide produces hydroxyl radicals, which are highly reactive and can oxidize contaminants. Hydroxyl radicals are used for water treatment in methods generally termed advanced oxidation processes. Because light is required for this reaction, TiO_2 is not appropriate for underground *in situ* remediation, but it may be used for wastewater treatment or pump-and-treat groundwater remediation.

TiO_2 is inexpensive, chemically stable, and insoluble in water. TiO_2 has a wide band gap energy (3.2 eV) that requires the use of UV light, as opposed to visible light only, for photocatalytic activation. To enhance the efficiency of its photocatalysis, research has investigated modifications to TiO_2 or alternative photocatalysts that might use a greater portion of photons in the visible light spectrum. Potential modifications include doping TiO_2 with metals, nitrogen, or carbon.

Challenges

When using *in situ* remediation the reactive products must be considered for two reasons. One reason is that a reactive product might be more harmful or mobile than the parent compound. Another reason is that the products can affect the effectiveness and cost of remediation. TCE (trichloroethylene), under reducing conditions by nanoiron, may sequentially dechlorinate to DCE (dichloroethene) and VC (vinyl chloride). VC is known to be more harmful than TCE, meaning this process would be undesirable.

Nanoparticles also react with non-target compounds. Bare nanoparticles tend to clump together and also react rapidly with soil, sediment, or other material in ground water. For *in situ* remediation, this action inhibits the particles from dispersing in the contaminated area, reducing their effectiveness for remediation. Coatings or other treatment may allow nanoparticles to disperse farther and potentially reach a greater portion of the contaminated zone. Coatings for nZVI include surfactants, polyelectrolyte coatings, emulsification layers, and protective shells made from silica or carbon.

Such designs may also affect the nanoparticles' ability to react with contaminants, their uptake by organisms, and their toxicity. A continuing area of research involves the potential for nanoparticles used for remediation to disperse widely and harm wildlife, plants, or people.

In some cases, bioremediation may be used deliberately at the same site or with the same material as nanoremediation. Ongoing research is investigating how nanoparticles may interact with simultaneous biological remediation.

DNA NANOTECHNOLOGY

DNA nanotechnology is the design and manufacture of artificial nucleic acid structures for technological uses. In this field, nucleic acids are used as non-biological engineering materials for nanotechnology rather than as the carriers of genetic information in living cells. Researchers in the field have created static structures such as two- and three-dimensional crystal lattices, nanotubes, polyhedra, and arbitrary shapes, and functional devices such as molecular machines and DNA computers. The field is beginning to be used as a tool to solve basic science problems in structural biology and biophysics, including applications in X-ray crystallography and nuclear magnetic resonance spectroscopy of proteins to determine structures. Potential applications in molecular scale electronics and nanomedicine are also being investigated.

The conceptual foundation for DNA nanotechnology was first laid out by Nadrian Seeman in the early 1980s, and the field began to attract widespread interest in the mid-2000s. This use of nucleic acids is enabled by their strict base pairing rules, which cause only portions of strands with complementary base sequences to bind together to form strong, rigid double helix structures. This allows for the rational design of base sequences that will selectively assemble to form complex target structures with precisely controlled nanoscale features. Several assembly methods are used to make these structures, including tile-based structures that assemble from smaller structures, folding structures using the DNA origami method, and dynamically reconfigurable structures using strand displacement methods. The field's name specifically references DNA, but the same principles have been used with other types of nucleic acids as well, leading to the occasional use of the alternative name *nucleic acid nanotechnology.*

Fundamental Concepts

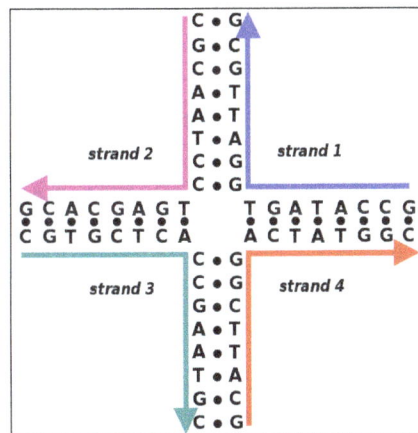

These four strands associate into a DNA four-arm junction because this structure maximizes the number of correct base pairs, with A matched to T and C matched to G.

This double-crossover (DX) supramolecular complex consists of five DNA single strands that form two double-helical domains, on the top and the bottom in this image. There are two crossover points where the strands cross from one domain into the other.

Properties of Nucleic Acids

Nanotechnology is often defined as the study of materials and devices with features on a scale below 100 nanometers. DNA nanotechnology, specifically, is an example of bottom-up molecular self-assembly, in which molecular components spontaneously organize into stable structures; the particular form of these structures is induced by the physical and chemical properties of the components selected by the designers. In DNA nanotechnology, the component materials are strands of nucleic acids such as DNA; these strands are often synthetic and are almost always used outside the context of a living cell. DNA is well-suited to nanoscale construction because the binding between two nucleic acid strands depends on simple base pairing rules which are well understood, and form the specific nanoscale structure of the nucleic acid double helix. These qualities make the assembly of nucleic acid structures easy to control through nucleic acid design. This property is absent in other materials used in nanotechnology, including proteins, for which protein design is very difficult, and nanoparticles, which lack the capability for specific assembly on their own.

The structure of a nucleic acid molecule consists of a sequence of nucleotides distinguished by which nucleobase they contain. In DNA, the four bases present are adenine (A), cytosine (C), guanine (G), and thymine (T). Nucleic acids have the property that two molecules will only bind to each other to form a double helix if the two sequences are complementary, meaning that they form matching sequences of base pairs, with A only binding to T, and C only to G. Because the formation of correctly matched base pairs is energetically favorable, nucleic acid strands are expected in most cases to bind to each other in the conformation that maximizes the number of correctly paired bases. The sequences of bases in a system of strands thus determine the pattern of binding and the overall structure in an easily controllable way. In DNA nanotechnology, the base sequences of strands are rationally designed by researchers so that the base pairing interactions cause the strands to assemble in the desired conformation. While DNA is the dominant material used, structures incorporating other nucleic acids such as RNA and peptide nucleic acid (PNA) have also been constructed.

Subfields

DNA nanotechnology is sometimes divided into two overlapping subfields: structural DNA nanotechnology and dynamic DNA nanotechnology. Structural DNA nanotechnology, sometimes abbreviated as SDN, focuses on synthesizing and characterizing nucleic acid complexes and materials that assemble into a static, equilibrium end state. On the other hand, dynamic DNA nanotechnology focuses on complexes with useful non-equilibrium behavior such as the ability to reconfigure based on a chemical or physical stimulus. Some complexes, such as nucleic acid nanomechanical devices, combine features of both the structural and dynamic subfields.

The complexes constructed in structural DNA nanotechnology use topologically branched nucleic acid structures containing junctions. (In contrast, most biological DNA exists as an unbranched double helix). One of the simplest branched structures is a four-arm junction that consists of four individual DNA strands, portions of which are complementary in a specific pattern. Unlike in natural Holliday junctions, each arm in the artificial immobile four-arm junction has a different base sequence, causing the junction point to be fixed at a certain position. Multiple junctions can be combined in the same complex, such as in the widely used double-crossover (DX) structural motif, which contains two parallel double helical domains with individual strands crossing between the domains at two crossover points. Each crossover point is, topologically, a four-arm junction, but is constrained to one orientation, in contrast to the flexible single four-arm junction, providing a rigidity that makes the DX motif suitable as a structural building block for larger DNA complexes.

Dynamic DNA nanotechnology uses a mechanism called toehold-mediated strand displacement to allow the nucleic acid complexes to reconfigure in response to the addition of a new nucleic acid strand. In this reaction, the incoming strand binds to a single-stranded toehold region of a double-stranded complex, and then displaces one of the strands bound in the original complex through a branch migration process. The overall effect is that one of the strands in the complex is replaced with another one. In addition, reconfigurable structures and devices can be made using functional nucleic acids such as deoxyribozymes and ribozymes, which can perform chemical reactions, and aptamers, which can bind to specific proteins or small molecules.

Structural DNA Nanotechnology

Structural DNA nanotechnology, sometimes abbreviated as SDN, focuses on synthesizing and characterizing nucleic acid complexes and materials where the assembly has a static, equilibrium endpoint. The nucleic acid double helix has a robust, defined three-dimensional geometry that makes it possible to predict and design the structures of more complicated nucleic acid complexes. Many such structures have been created, including two- and three-dimensional structures, and periodic, aperiodic, and discrete structures.

Extended Lattices

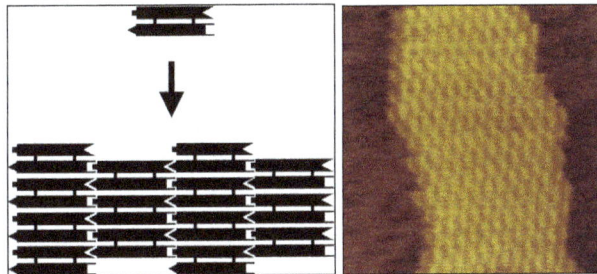

In figure, the assembly of a DX array. *Left*, schematic diagram. Each bar represents a double-helical domain of DNA, with the shapes representing complementary sticky ends. The DX complex at top will combine with other DX complexes into the two-dimensional array shown at bottom. *Right*, an atomic force microscopy image of the assembled array. The individual DX tiles are clearly visible within the assembled structure. The field is 150 nm across.

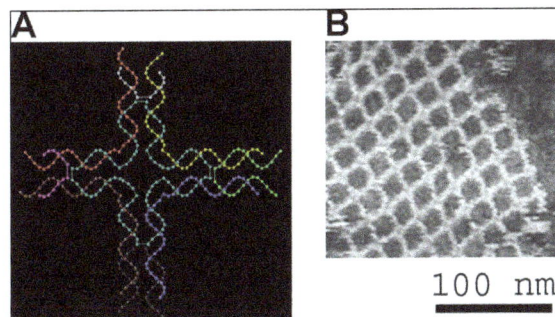

Left, a model of a DNA tile used to make another two-dimensional periodic lattice.
Right, an atomic force micrograph of the assembled lattice.

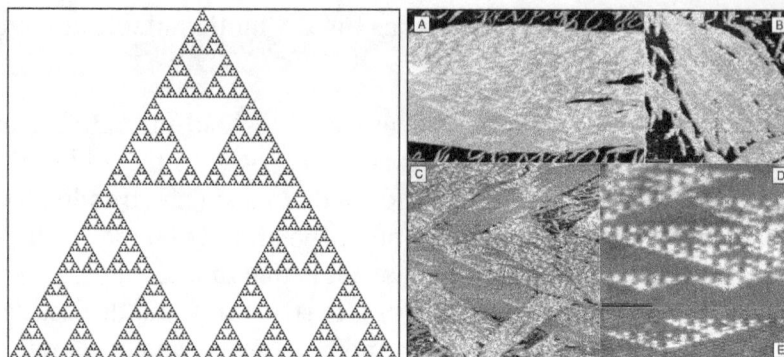

An example of an aperiodic two-dimensional lattice that assembles into a fractal pattern. *Left*, the Sierpinski gasket fractal. *Right*, DNA arrays that display a representation of the Sierpinski gasket on their surfaces.

Small nucleic acid complexes can be equipped with sticky ends and combined into larger two-dimensional periodic lattices containing a specific tessellated pattern of the individual molecular tiles. The earliest example of this used double-crossover (DX) complexes as the basic tiles, each containing four sticky ends designed with sequences that caused the DX units to combine into periodic two-dimensional flat sheets that are essentially rigid two-dimensional crystals of DNA. Two-dimensional arrays have been made from other motifs as well, including the Holliday junction rhombus lattice, and various DX-based arrays making use of a double-cohesion scheme. The top two images at right show examples of tile-based periodic lattices.

Two-dimensional arrays can be made to exhibit aperiodic structures whose assembly implements a specific algorithm, exhibiting one form of DNA computing. The DX tiles can have their sticky end sequences chosen so that they act as Wang tiles, allowing them to perform computation. A DX array whose assembly encodes an XOR operation has been demonstrated; this allows the DNA array to implement a cellular automaton that generates a fractal known as the Sierpinski gasket. The third image at right shows this type of array. Another system has the function of a binary counter, displaying a representation of increasing binary numbers as it grows. These results show that computation can be incorporated into the assembly of DNA arrays.

DX arrays have been made to form hollow nanotubes 4–20 nm in diameter, essentially two-dimensional lattices which curve back upon themselves. These DNA nanotubes are somewhat similar in size and shape to carbon nanotubes, and while they lack the electrical conductance of carbon nanotubes, DNA nanotubes are more easily modified and connected to other structures. One of many schemes for constructing DNA nanotubes uses a lattice of curved DX tiles that curls around itself and closes into a tube. In an alternative method that allows the circumference to be specified in a simple, modular fashion using single-stranded tiles, the rigidity of the tube is an emergent property.

Forming three-dimensional lattices of DNA was the earliest goal of DNA nanotechnology, but this proved to be one of the most difficult to realize. Success using a motif based on the concept of tensegrity, a balance between tension and compression forces, was finally reported in 2009.

Discrete Structures

Researchers have synthesized many three-dimensional DNA complexes that each have the connectivity of a polyhedron, such as a cube or octahedron, meaning that the DNA duplexes trace the edges of a polyhedron with a DNA junction at each vertex. The earliest demonstrations of DNA polyhedra were very work-intensive, requiring multiple ligations and solid-phase synthesis steps to create catenated polyhedra. Subsequent work yielded polyhedra whose synthesis was much easier. These include a DNA octahedron made from a long single strand designed to fold into the correct conformation, and a tetrahedron that can be produced from four DNA strands in one step.

Nanostructures of arbitrary, non-regular shapes are usually made using the DNA origami method. These structures consist of a long, natural virus strand as a "scaffold", which is made to fold into the desired shape by computationally designed short "staple" strands. This method has the advantages of being easy to design, as the base sequence is predetermined by the scaffold strand sequence, and not requiring high strand purity and accurate stoichiometry, as most other DNA nanotechnology methods do. DNA origami was first demonstrated for two-dimensional shapes, such as a smiley face, a coarse map of the Western Hemisphere, and the Mona Lisa painting. Solid three-dimensional structures can be made by using parallel DNA helices arranged in a honeycomb pattern, and structures with two-dimensional faces can be made to fold into a hollow overall three-dimensional shape, akin to a cardboard box. These can be programmed to open and reveal or release a molecular cargo in response to a stimulus, making them potentially useful as programmable molecular cages.

Templated Assembly

Nucleic acid structures can be made to incorporate molecules other than nucleic acids, sometimes called heteroelements, including proteins, metallic nanoparticles, quantum dots, and fullerenes. This allows the construction of materials and devices with a range of functionalities much greater than is possible with nucleic acids alone. The goal is to use the self-assembly of the nucleic acid structures to template the assembly of the nanoparticles hosted on them, controlling their position and in some cases orientation. Many of these schemes use a covalent attachment scheme, using oligonucleotides with amide or thiol functional groups as a chemical handle to bind the heteroelements. This covalent binding scheme has been used to arrange gold nanoparticles on a DX-based array, and to arrange streptavidin protein molecules into specific patterns on a DX array. A non-covalent hosting scheme using Dervan polyamides on a DX array was used to arrange streptavidin proteins in a specific pattern on a DX array. Carbon nanotubes have been hosted on DNA arrays in a pattern allowing the assembly to act as a molecular electronic device, a carbon nanotube field-effect transistor. In addition, there are nucleic acid metallization methods, in which the nucleic acid is replaced by a metal which assumes the general shape of the original nucleic acid structure, and schemes for using nucleic acid nanostructures as lithography masks, transferring their pattern into a solid surface.

Dynamic DNA Nanotechnology

Dynamic DNA nanotechnology focuses on forming nucleic acid systems with designed dynamic functionalities related to their overall structures, such as computation and mechanical motion. There is some overlap between structural and dynamic DNA nanotechnology, as structures can be formed through annealing and then reconfigured dynamically, or can be made to form dynamically in the first place.

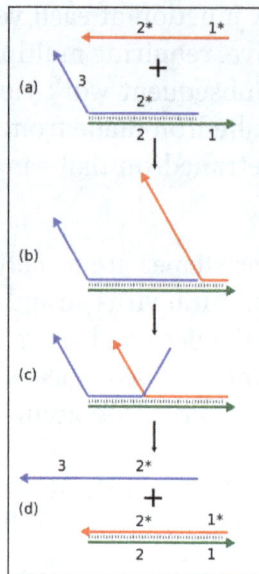

Dynamic DNA nanotechnology often makes use of toehold-mediated strand displacement reactions. In this example, the red strand binds to the single stranded toehold region on the green strand (region 1), and then in a branch migration process across region 2, the blue strand is displaced and freed from the complex. Reactions like these are used to dynamically reconfigure or assemble nucleic acid nanostructures. In addition, the red and blue strands can be used as signals in a molecular logic gate.

Nanomechanical Devices

DNA complexes have been made that change their conformation upon some stimulus, making them one form of nanorobotics. These structures are initially formed in the same way as the static structures made in structural DNA nanotechnology, but are designed so that dynamic reconfiguration is possible after the initial assembly. The earliest such device made use of the transition between the B-DNA and Z-DNA forms to respond to a change in buffer conditions by undergoing a twisting motion. This reliance on buffer conditions caused all devices to change state at the same time. Subsequent systems could change states based upon the presence of control strands, allowing multiple devices to be independently operated in solution. Some examples of such systems are a "molecular tweezers" design that has an open and a closed state, a device that could switch from a paranemic-crossover (PX) conformation to a double-junction (JX2) conformation, undergoing rotational motion in the process, and a two-dimensional array that could dynamically expand and contract in response to control strands. Structures have also been made that dynamically open or close, potentially acting as a molecular cage to release or reveal a functional cargo upon opening.

DNA walkers are a class of nucleic acid nanomachines that exhibit directional motion along a linear track. A large number of schemes have been demonstrated. One strategy is to control the motion of the walker along the track using control strands that need to be manually added in sequence. Another approach is to make use of restriction enzymes or deoxyribozymes to cleave the strands and cause the walker to move forward, which has the advantage of running autonomously. A later system could walk upon a two-dimensional surface rather than a linear track, and demonstrated the ability to selectively pick up and move molecular cargo. Additionally, a linear walker has been demonstrated that performs DNA-templated synthesis as the walker advances along the track, allowing autonomous multistep chemical synthesis directed by the walker. The synthetic DNA walkers' function is similar to that of the proteins dynein and kinesin.

Strand Displacement Cascades

Cascades of strand displacement reactions can be used for either computational or structural purposes. An individual strand displacement reaction involves revealing a new sequence in response to the presence of some initiator strand. Many such reactions can be linked into a cascade where the newly revealed output sequence of one reaction can initiate another strand displacement reaction elsewhere. This in turn allows for the construction of chemical reaction networks with many components, exhibiting complex computational and information processing abilities. These cascades are made energetically favorable through the formation of new base pairs, and the entropy gain from disassembly reactions. Strand displacement cascades allow isothermal operation of the assembly or computational process, in contrast to traditional nucleic acid assembly's requirement for a thermal annealing step, where the temperature is raised and then slowly lowered to ensure proper formation of the desired structure. They can also support catalytic function of the initiator species, where less than one equivalent of the initiator can cause the reaction to go to completion.

Strand displacement complexes can be used to make molecular logic gates capable of complex computation. Unlike traditional electronic computers, which use electric current as inputs and outputs, molecular computers use the concentrations of specific chemical species as signals. In the case of nucleic acid strand displacement circuits, the signal is the presence of nucleic acid strands that are released or consumed by binding and unbinding events to other strands in displacement complexes. This approach has been used to make logic gates such as AND, OR, and NOT gates. More recently, a four-bit circuit was demonstrated that can compute the square root of the integers 0–15, using a system of gates containing 130 DNA strands.

Another use of strand displacement cascades is to make dynamically assembled structures. These use a hairpin structure for the reactants, so that when the input strand binds, the newly revealed sequence is on the same molecule rather than disassembling. This allows new opened hairpins to be added to a growing complex. This approach has been used to make simple structures such as three- and four-arm junctions and dendrimers.

Applications

DNA nanotechnology provides one of the few ways to form designed, complex structures with precise control over nanoscale features. The field is beginning to see application to solve basic science problems in structural biology and biophysics. The earliest such application envisaged for the field, and one still in development, is in crystallography, where molecules that are difficult to crystallize in isolation could be arranged within a three-dimensional nucleic acid lattice, allowing determination of their structure. Another application is the use of DNA origami rods to replace liquid crystals in residual dipolar coupling experiments in protein NMR spectroscopy; using DNA origami is advantageous because, unlike liquid crystals, they are tolerant of the detergents needed to suspend membrane proteins in solution. DNA walkers have been used as nanoscale assembly lines to move nanoparticles and direct chemical synthesis. Further, DNA origami structures have aided in the biophysical studies of enzyme function and protein folding.

DNA nanotechnology is moving toward potential real-world applications. The ability of nucleic acid arrays to arrange other molecules indicates its potential applications in molecular scale electronics. The assembly of a nucleic acid structure could be used to template the assembly of a molecular electronic elements such as molecular wires, providing a method for nanometer-scale control of the placement and overall architecture of the device analogous to a molecular breadboard. DNA nanotechnology has been compared to the concept of programmable matter because of the coupling of computation to its material properties.

In a study conducted by a group of scientists from iNANO and CDNA centers in Aarhus University, researchers were able to construct a small multi-switchable 3D DNA Box Origami. The proposed nanoparticle was characterized by atomic force microscopy (AFM), transmission electron microscopy (TEM) and Förster resonance energy transfer (FRET). The constructed box was shown to have a unique reclosing mechanism, which enabled it to repeatedly open and close in response to a unique set of DNA or RNA keys. The authors proposed that this "DNA device can potentially be used for a broad range of applications such as controlling the function of single molecules, controlled drug delivery, and molecular computing."

There are potential applications for DNA nanotechnology in nanomedicine, making use of its ability to perform computation in a biocompatible format to make "smart drugs" for targeted drug delivery. One such system being investigated uses a hollow DNA box containing proteins that induce apoptosis, or cell death, that will only open when in proximity to a cancer cell. There has additionally been interest in expressing these artificial structures in engineered living bacterial cells, most likely using the transcribed RNA for the assembly, although it is unknown whether these complex structures are able to efficiently fold or assemble in the cell's cytoplasm. If successful, this could enable directed evolution of nucleic acid nanostructures. Scientists at Oxford University reported the self-assembly of four short strands of synthetic DNA into a cage which can enter cells and survive for at least 48 hours. The fluorescently labeled DNA tetrahedra were found to remain intact in the laboratory cultured human kidney cells despite the attack by cellular enzymes after two days. This experiment showed the potential of drug delivery inside the living cells using the DNA 'cage'. A DNA tetrahedron was used to deliver RNA Interference (RNAi) in a mouse model, reported a team of researchers in MIT. Delivery of the interfering RNA for treatment has showed some success using polymer or lipid, but there are limits of safety and imprecise targeting, in addition to short shelf life in the blood stream. The DNA nanostructure created by the team consists of six strands of DNA to form a tetrahedron, with one strand of RNA affixed to each of the six edges. The tetrahedron is further equipped with targeting protein, three folate molecules, which lead the DNA nanoparticles to the abundant folate receptors found on some tumors. The result showed that the gene expression targeted by the RNAi, luciferase, dropped by more than half. This study shows promise in using DNA nanotechnology as an effective tool to deliver treatment using the emerging RNA Interference technology. The DNA tetrahedron was also used in an effort to overcome the phenomena multidrug resistance. Doxorubicin (DOX) was conjugated with the tetrahedron and was loaded into MCF-7 breast cancer cells that contained the P-glycoprotein drug efflux pump. The results of the experiment showed the DOX was not being pumped out and apoptosis of the cancer cells was achieved. The tetrahedron without DOX was loaded into cells to test its biocompatibility, and the structure showed no cytotoxicity itself.

Applications for DNA nanotechnology in nanomedicine also focus on mimicking the structure and function of naturally occurring membrane proteins with designed DNA nanostructures. In

2012, Langecker *et al.* introduced a pore-shaped DNA origami structure that can self-insert into lipid membranes via hydrophobic cholesterol modifications and induce ionic currents across the membrane. This first demonstration of a synthetic DNA ion channel was followed by a variety of pore-inducing designs ranging from a single DNA duplex, to small tile-based structures, and large DNA origami transmembrane porins. Similar to naturally occurring protein ion channels, this ensemble of synthetic DNA-made counterparts thereby spans multiple orders of magnitude in conductance. The study of the membrane-inserting single DNA duplex showed that current must also flow on the DNA-lipid interface as no central channel lumen is present in the design that lets ions pass across the lipid bilayer. This indicated that the DNA-induced lipid pore has a toroidal shape, rather than cylindrical, as lipid headgroups reorient to face towards the membrane-inserted part of the DNA. Researchers from the University of Cambridge and the University of Illinois at Urbana-Champaign then demonstrated that such a DNA-induced toroidal pore can facilitate rapid lipid flip-flop between the lipid bilayer leaflets. Utilizing this effect, they designed a synthetic DNA-built enzyme that flips lipids in biological membranes orders of magnitudes faster than naturally occurring proteins called scramblases. This development highlights the potential of synthetic DNA nanostructures for personalized drugs and therapeutics.

Design

DNA nanostructures must be rationally designed so that individual nucleic acid strands will assemble into the desired structures. This process usually begins with specification of a desired target structure or function. Then, the overall secondary structure of the target complex is determined, specifying the arrangement of nucleic acid strands within the structure, and which portions of those strands should be bound to each other. The last step is the primary structure design, which is the specification of the actual base sequences of each nucleic acid strand.

Structural Design

The first step in designing a nucleic acid nanostructure is to decide how a given structure should be represented by a specific arrangement of nucleic acid strands. This design step determines the secondary structure, or the positions of the base pairs that hold the individual strands together in the desired shape. Several approaches have been demonstrated:

- Tile-based structures: This approach breaks the target structure into smaller units with strong binding between the strands contained in each unit, and weaker interactions between the units. It is often used to make periodic lattices, but can also be used to implement algorithmic self-assembly, making them a platform for DNA computing. This was the dominant design strategy used from the mid-1990s until the mid-2000s, when the DNA origami methodology was developed.

- Folding structures: An alternative to the tile-based approach, folding approaches make the nanostructure from one long strand, which can either have a designed sequence that folds due to its interactions with itself, or it can be folded into the desired shape by using shorter, "staple" strands. This latter method is called DNA origami, which allows forming nanoscale two- and three-dimensional shapes.

- Dynamic assembly: This approach directly controls the kinetics of DNA self-assembly, specifying all of the intermediate steps in the reaction mechanism in addition to the final product. This is done using starting materials which adopt a hairpin structure; these then assemble into the final conformation in a cascade reaction, in a specific order. This approach has the advantage of proceeding isothermally, at a constant temperature. This is in contrast to the thermodynamic approaches, which require a thermal annealing step where a temperature change is required to trigger the assembly and favor proper formation of the desired structure.

Sequence Design

After any of the above approaches are used to design the secondary structure of a target complex, an actual sequence of nucleotides that will form into the desired structure must be devised. Nucleic acid design is the process of assigning a specific nucleic acid base sequence to each of a structure's constituent strands so that they will associate into a desired conformation. Most methods have the goal of designing sequences so that the target structure has the lowest energy, and is thus the most thermodynamically favorable, while incorrectly assembled structures have higher energies and are thus disfavored. This is done either through simple, faster heuristic methods such as sequence symmetry minimization, or by using a full nearest-neighbor thermodynamic model, which is more accurate but slower and more computationally intensive. Geometric models are used to examine tertiary structure of the nanostructures and to ensure that the complexes are not overly strained.

Nucleic acid design has similar goals to protein design. In both, the sequence of monomers is designed to favor the desired target structure and to disfavor other structures. Nucleic acid design has the advantage of being much computationally easier than protein design, because the simple base pairing rules are sufficient to predict a structure's energetic favorability, and detailed information about the overall three-dimensional folding of the structure is not required. This allows the use of simple heuristic methods that yield experimentally robust designs. Nucleic acid structures are less versatile than proteins in their function because of proteins' increased ability to fold into complex structures, and the limited chemical diversity of the four nucleotides as compared to the twenty proteinogenic amino acids.

Materials and Methods

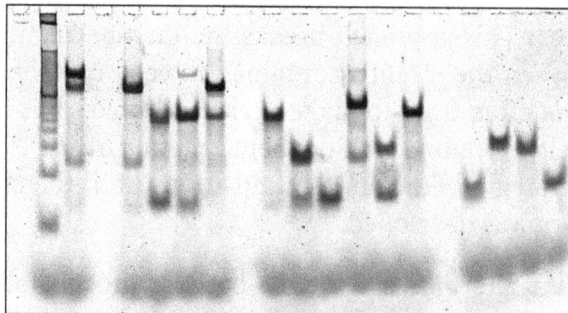

Gel electrophoresis methods, such as this formation assay on a DX complex, are used to ascertain whether the desired structures are forming properly. Each vertical lane contains a series of bands, where each band is characteristic of a particular reaction intermediate.

The sequences of the DNA strands making up a target structure are designed computationally, using molecular modeling and thermodynamic modeling software. The nucleic acids themselves are then synthesized using standard oligonucleotide synthesis methods, usually automated in an oligonucleotide synthesizer, and strands of custom sequences are commercially available. Strands can be purified by denaturing gel electrophoresis if needed, and precise concentrations determined via any of several nucleic acid quantitation methods using ultraviolet absorbance spectroscopy.

The fully formed target structures can be verified using native gel electrophoresis, which gives size and shape information for the nucleic acid complexes. An electrophoretic mobility shift assay can assess whether a structure incorporates all desired strands. Fluorescent labeling and Förster resonance energy transfer (FRET) are sometimes used to characterize the structure of the complexes.

Nucleic acid structures can be directly imaged by atomic force microscopy, which is well suited to extended two-dimensional structures, but less useful for discrete three-dimensional structures because of the microscope tip's interaction with the fragile nucleic acid structure; transmission electron microscopy and cryo-electron microscopy are often used in this case. Extended three-dimensional lattices are analyzed by X-ray crystallography.

NANOMEDICINE

Nanomedicine is the medical application of nanotechnology. Nanomedicine ranges from the medical applications of nanomaterials and biological devices, to nanoelectronic biosensors, and even possible future applications of molecular nanotechnology such as biological machines. Current problems for nanomedicine involve understanding the issues related to toxicity and environmental impact of nanoscale materials (materials whose structure is on the scale of nanometers, i.e. billionths of a meter).

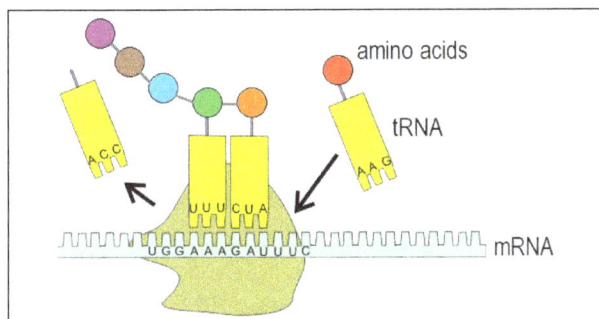

A ribosome is a biological machine.

Functionalities can be added to nanomaterials by interfacing them with biological molecules or structures. The size of nanomaterials is similar to that of most biological molecules and structures; therefore, nanomaterials can be useful for both in vivo and in vitro biomedical research and applications. Thus far, the integration of nanomaterials with biology has led to the development of diagnostic devices, contrast agents, analytical tools, physical therapy applications, and drug delivery vehicles.

Nanomedicine seeks to deliver a valuable set of research tools and clinically useful devices in the near future. The National Nanotechnology Initiative expects new commercial applications in the pharmaceutical industry that may include advanced drug delivery systems, new therapies, and in vivo imaging. Nanomedicine research is receiving funding from the US National Institutes of Health Common Fund program, supporting four nanomedicine development centers.

Nanomedicine sales reached $16 billion in 2015, with a minimum of $3.8 billion in nanotechnology R&D being invested every year. Global funding for emerging nanotechnology increased by 45% per year in recent years, with product sales exceeding $1 trillion in 2013. As the nanomedicine industry continues to grow, it is expected to have a significant impact on the economy.

Drug Delivery

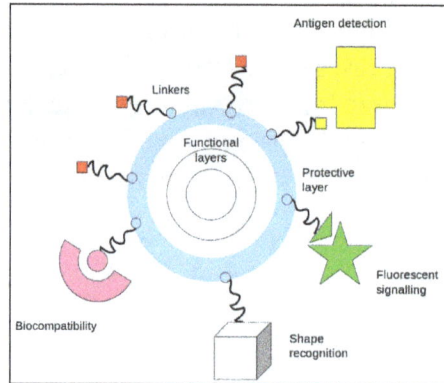

Nanotechnology has provided the possibility of delivering drugs to specific cells using nanoparticles. The overall drug consumption and side-effects may be lowered significantly by depositing the active agent in the morbid region only and in no higher dose than needed. Targeted drug delivery is intended to reduce the side effects of drugs with concomitant decreases in consumption and treatment expenses. Drug delivery focuses on maximizing bioavailability both at specific places in the body and over a period of time. This can potentially be achieved by molecular targeting by nanoengineered devices. A benefit of using nanoscale for medical technologies is that smaller devices are less invasive and can possibly be implanted inside the body, plus biochemical reaction times are much shorter. These devices are faster and more sensitive than typical drug delivery. The efficacy of drug delivery through nanomedicine is largely based upon: a) efficient encapsulation of the drugs, b) successful delivery of drug to the targeted region of the body, and c) successful release of the drug.

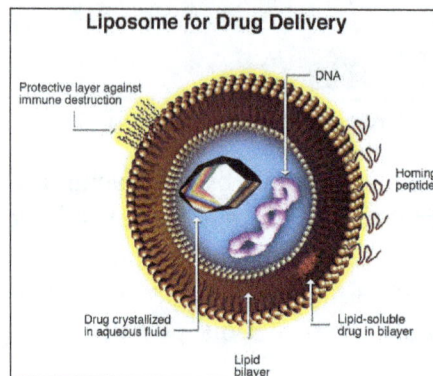

Drug delivery systems, lipid- or polymer-based nanoparticles, can be designed to improve the pharmacokinetics and biodistribution of the drug. However, the pharmacokinetics and pharmacodynamics of nanomedicine is highly variable among different patients. When designed to avoid the body's defence mechanisms, nanoparticles have beneficial properties that can be used to improve drug delivery. Complex drug delivery mechanisms are being developed, including the ability to get drugs through cell membranes and into cell cytoplasm. Triggered response is one way for drug molecules to be used more efficiently. Drugs are placed in the body and only activate on encountering a particular signal. For example, a drug with poor solubility will be replaced by a drug delivery system where both hydrophilic and hydrophobic environments exist, improving the solubility. Drug delivery systems may also be able to prevent tissue damage through regulated drug release; reduce drug clearance rates; or lower the volume of distribution and reduce the effect on non-target tissue. However, the biodistribution of these nanoparticles is still imperfect due to the complex host's reactions to nano- and microsized materials and the difficulty in targeting specific organs in the body. Nevertheless, a lot of work is still ongoing to optimize and better understand the potential and limitations of nanoparticulate systems. While advancement of research proves that targeting and distribution can be augmented by nanoparticles, the dangers of nanotoxicity become an important next step in further understanding of their medical uses.

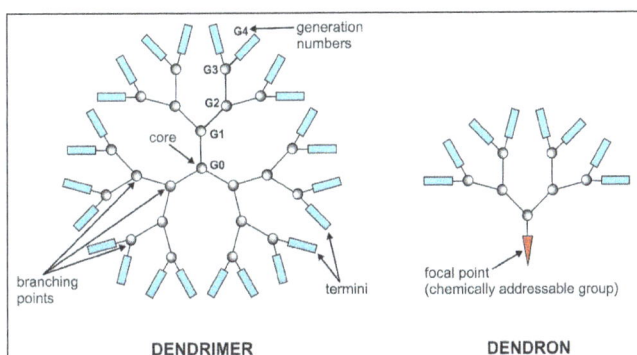

Nanoparticles (top), liposomes (middle), and dendrimers (bottom) are some nanomaterials being investigated for use in nanomedicine.

Nanoparticles are under research for their potential to decrease antibiotic resistance or for various antimicrobial uses. Nanoparticles might also be used to circumvent multidrug resistance (MDR) mechanisms.

Systems under Research

Advances in lipid nanotechnology were instrumental in engineering medical nanodevices and novel drug delivery systems, as well as in developing sensing applications. Another system for microRNA delivery under preliminary research is nanoparticles formed by the self-assembly of two different microRNAs deregulated in cancer. One potential application is based on small electromechanical systems, such as nanoelectromechanical systems being investigated for the active release of drugs and sensors for possible cancer treatment with iron nanoparticles or gold shells.

Applications

Some nanotechnology-based drugs that are commercially available or in human clinical trials include:

- Abraxane, approved by the U.S. Food and Drug Administration (FDA) to treat breast cancer, non-small-cell lung cancer (NSCLC) and pancreatic cancer, is the nanoparticle albumin bound paclitaxel.

- Doxil was originally approved by the FDA for the use on HIV-related Kaposi's sarcoma. It is now being used to also treat ovarian cancer and multiple myeloma. The drug is encased in liposomes, which helps to extend the life of the drug that is being distributed. Liposomes are self-assembling, spherical, closed colloidal structures that are composed of lipid bilayers that surround an aqueous space. The liposomes also help to increase the functionality and it helps to decrease the damage that the drug does to the heart muscles specifically.

- Onivyde, liposome encapsulated irinotecan to treat metastatic pancreatic cancer, was approved by FDA in October 2015.

- Rapamune is a nanocrystal-based drug that was approved by the FDA in 2000 to prevent organ rejection after transplantation. The nanocrystal components allow for increased drug solubility and dissolution rate, leading to improved absorption and high bioavailability.

Cancer

Preclinical Research

Existing and potential drug nanocarriers have been reviewed.

Nanoparticles have high surface area to volume ratio. This allows for many functional groups to be attached to a nanoparticle, which can seek out and bind to certain tumor cells. Additionally, the small size of nanoparticles (5 to 100 nanometers), allows them to preferentially accumulate at tumor sites (because tumors lack an effective lymphatic drainage system). Limitations to conventional cancer chemotherapy include drug resistance, lack of selectivity, and lack of solubility.

Imaging

In vivo imaging is another area where tools and devices are being developed. Using nanoparticle contrast agents, images such as ultrasound and MRI have a favorable distribution and improved contrast. In cardiovascular imaging, nanoparticles have potential to aid visualization of blood pooling, ischemia, angiogenesis, atherosclerosis, and focal areas where inflammation is present.

The small size of nanoparticles endows them with properties that can be very useful in oncology, particularly in imaging. Quantum dots (nanoparticles with quantum confinement properties, such as size-tunable light emission), when used in conjunction with MRI (magnetic resonance imaging), can produce exceptional images of tumor sites. Nanoparticles of cadmium selenide (quantum dots) glow when exposed to ultraviolet light. When injected, they seep into cancer tumors. The surgeon can see the glowing tumor, and use it as a guide for more accurate tumor removal.These nanoparticles are much brighter than organic dyes and only need one light source for excitation. This means that the use of fluorescent quantum dots could produce a higher contrast image and at a lower cost than today's organic dyes used as contrast media. The downside, however, is that quantum dots are usually made of quite toxic elements, but this concern may be addressed by use of fluorescent dopants.

Tracking movement can help determine how well drugs are being distributed or how substances are metabolized. It is difficult to track a small group of cells throughout the body, so scientists used to dye the cells. These dyes needed to be excited by light of a certain wavelength in order for them to light up. While different color dyes absorb different frequencies of light, there was a need for as many light sources as cells. A way around this problem is with luminescent tags. These tags are quantum dots attached to proteins that penetrate cell membranes. The dots can be random in size, can be made of bio-inert material, and they demonstrate the nanoscale property that color is size-dependent. As a result, sizes are selected so that the frequency of light used to make a group of quantum dots fluoresce is an even multiple of the frequency required to make another group incandesce. Then both groups can be lit with a single light source. They have also found a way to insert nanoparticles into the affected parts of the body so that those parts of the body will glow showing the tumor growth or shrinkage or also organ trouble.

Sensing

Nanotechnology-on-a-chip is one more dimension of lab-on-a-chip technology. Magnetic nanoparticles, bound to a suitable antibody, are used to label specific molecules, structures or microorganisms. In particular silica nanoparticles are inert from the photophysical point of view and might accumulate a large number of dyes within the nanoparticle shell. Gold nanoparticles tagged with short segments of DNA can be used for detection of genetic sequence in a sample. Multicolor optical coding for biological assays has been achieved by embedding different-sized quantum dots into polymeric microbeads. Nanopore technology for analysis of nucleic acids converts strings of nucleotides directly into electronic signatures.

Sensor test chips containing thousands of nanowires, able to detect proteins and other biomarkers left behind by cancer cells, could enable the detection and diagnosis of cancer in the early stages from a few drops of a patient's blood. Nanotechnology is helping to advance the use of arthroscopes, which are pencil-sized devices that are used in surgeries with lights and cameras so surgeons can do the surgeries with smaller incisions. The smaller the incisions the faster the healing time which is better for the patients. It is also helping to find a way to make an arthroscope smaller than a strand of hair.

Research on nanoelectronics-based cancer diagnostics could lead to tests that can be done in pharmacies. The results promise to be highly accurate and the product promises to be inexpensive. They could take a very small amount of blood and detect cancer anywhere in the body in about five minutes, with a sensitivity that is a thousand times better a conventional laboratory test. These devices that are built with nanowires to detect cancer proteins; each nanowire detector is primed to be sensitive to a different cancer marker. The biggest advantage of the nanowire detectors is that they could test for anywhere from ten to one hundred similar medical conditions without adding cost to the testing device. Nanotechnology has also helped to personalize oncology for the detection, diagnosis, and treatment of cancer. It is now able to be tailored to each individual's tumor for better performance. They have found ways that they will be able to target a specific part of the body that is being affected by cancer.

Blood Purification

Magnetic micro particles are proven research instruments for the separation of cells and proteins from complex media. The technology is available under the name Magnetic-activated cell sorting or Dynabeads among others. More recently it was shown in animal models that magnetic nanoparticles can be used for the removal of various noxious compounds including toxins, pathogens, and proteins from whole blood in an extracorporeal circuit similar to dialysis. In contrast to dialysis, which works on the principle of the size related diffusion of solutes and ultrafiltration of fluid across a semi-permeable membrane, the purification with nanoparticles allows specific targeting of substances. Additionally larger compounds which are commonly not dialyzable can be removed.

The purification process is based on functionalized iron oxide or carbon coated metal nanoparticles with ferromagnetic or superparamagnetic properties. Binding agents such as proteins, antibodies, antibiotics, or synthetic ligands are covalently linked to the particle surface. These binding agents are able to interact with target species forming an agglomerate. Applying an external magnetic field gradient allows exerting a force on the nanoparticles. Hence the particles can be separated from the bulk fluid, thereby cleaning it from the contaminants.

The small size (< 100 nm) and large surface area of functionalized nanomagnets leads to advantageous properties compared to hemoperfusion, which is a clinically used technique for the purification of blood and is based on surface adsorption. These advantages are high loading and accessible for binding agents, high selectivity towards the target compound, fast diffusion, small hydrodynamic resistance, and low dosage.

This approach offers new therapeutic possibilities for the treatment of systemic infections such as sepsis by directly removing the pathogen. It can also be used to selectively remove cytokines or endotoxins or for the dialysis of compounds which are not accessible by traditional dialysis methods. However the technology is still in a preclinical phase and first clinical trials are not expected before 2017.

Tissue Engineering

Nanotechnology may be used as part of tissue engineering to help reproduce or repair or reshape damaged tissue using suitable nanomaterial-based scaffolds and growth factors. Tissue engineering if successful may replace conventional treatments like organ transplants or artificial implants. Nanoparticles such as graphene, carbon nanotubes, molybdenum disulfide and tungsten disulfide are being used as reinforcing agents to fabricate mechanically strong biodegradable polymeric nanocomposites for bone tissue engineering applications. The addition of these nanoparticles in the polymer matrix at low concentrations (~0.2 weight %) leads to significant improvements in the compressive and flexural mechanical properties of polymeric nanocomposites. Potentially, these nanocomposites may be used as a novel, mechanically strong, light weight composite as bone implants.

For example, a flesh welder was demonstrated to fuse two pieces of chicken meat into a single piece using a suspension of gold-coated nanoshells activated by an infrared laser. This could be used to weld arteries during surgery. Another example is nanonephrology, the use of nanomedicine on the kidney.

Medical Devices

Neuro-electronic interfacing is a visionary goal dealing with the construction of nanodevices that will permit computers to be joined and linked to the nervous system. This idea requires the building of a molecular structure that will permit control and detection of nerve impulses by an external computer. A refuelable strategy implies energy is refilled continuously or periodically with external sonic, chemical, tethered, magnetic, or biological electrical sources, while a nonrefuelable strategy implies that all power is drawn from internal energy storage which would stop when all energy is drained. A nanoscale enzymatic biofuel cell for self-powered nanodevices have been developed that uses glucose from biofluids including human blood and watermelons. One limitation to this innovation is the fact that electrical interference or leakage or overheating from power consumption is possible. The wiring of the structure is extremely difficult because they must be positioned precisely in the nervous system. The structures that will provide the interface must also be compatible with the body's immune system.

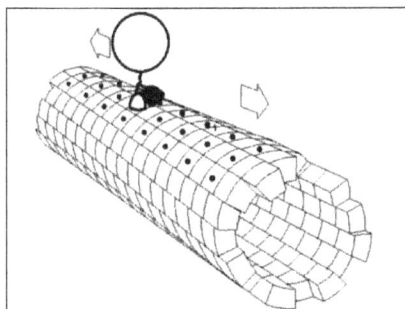

Kinesin is a protein complex functioning as a molecular biological machine.
It uses protein domain dynamics on nanoscales.

Molecular nanotechnology is a speculative subfield of nanotechnology regarding the possibility of engineering molecular assemblers, machines which could re-order matter at a molecular or atomic scale. Nanomedicine would make use of these nanorobots, introduced into the body, to repair or detect damages and infections. Molecular nanotechnology is highly theoretical, seeking to anticipate what inventions nanotechnology might yield and to propose an agenda for future inquiry. The proposed elements of molecular nanotechnology, such as molecular assemblers and nanorobots are far beyond current capabilities. Future advances in nanomedicine could give rise to life extension through the repair of many processes thought to be responsible for aging.

NANOTOXICOLOGY

Nanotoxicology is a branch of bionanoscience, which deals with the study and application of the toxicity of nanomaterials. Nanomaterials, even when made of inert elements such as gold, become highly active at nanometer dimensions. Nanotoxicological studies are used to determine whether and to what extent these properties may pose a threat to the environment and to human health.

Aquatic ecosystems are progressively coming under pressure due to the presence of emerging anthropogenic contaminants, including NMs, posing health hazards to inhabitant organisms. In recent years, increasing data demonstrated that NPs could induce toxicity and genotoxicity under a variety of exposure scenarios. An accepted mechanism by which NPs may induce cytotoxicity is considered to be through the induction of reactive oxygen species (ROS), which can induce oxidative stress, which in turn may lead to cytotoxicity, DNA damage, and other effects.

Physical and Chemical Properties of NMs Influencing their Toxicity

The behavior of NPs in various environmental matrices is complex and involves several processes. Properties of NMs are unique and different from conventional materials. Properties such as

(1) particle size, (2) surface area and charge, (3) shape/structure, (4) solubility, and (5) surface coatings are known to affect NP toxicity.

Owing to their small size, NMs have unique physical and chemical characteristics such as magnetic, optical, thermal, mechanical, electrical properties which make them suitable in several applications including in medicine, electronics, and energy production, and in several consumer products. However, these very properties have the potential to affect humans and the environment adversely. NPs can easily penetrate cell membranes and other biological barriers into living organisms causing cell damage. Studies reporting increased toxicity of NPs when compared to their larger bulk particles have led to a generally assumed hypothesis that NPs are more potent in causing damage.

Routes of Exposure in the Aquatic Environment

Due to the surge in nanotechnology, there have been significant increases in the number of various NPs released into the aquatic environment. Figure provides a summary of the possible routes in a typical aquatic environment to nanoparticles, potential interactions, and the possible clearance routes. Aquatic ecosystems are susceptible to environmental contamination since they are at the receiving end of contaminants, particularly from runoff sources. Identified sources of NPs in the aquatic environment include production facilities, production processes, wastewater treatment plants, and accidents during the transport. In addition, aquatic ecosystems are known to sequester and transport contaminants, including NMs. In the aquatic environment, NPs may aggregate thus reducing the NPs available for direct uptake in the aqueous phase by aquatic organisms. However, aggregated NPs may settle into sediment thereby posing a threat to benthic organisms. In the aquatic environment, NMs are generally associated with sediments. Sediments and soil represent porous environmental matrices which typically have large specific surface areas.

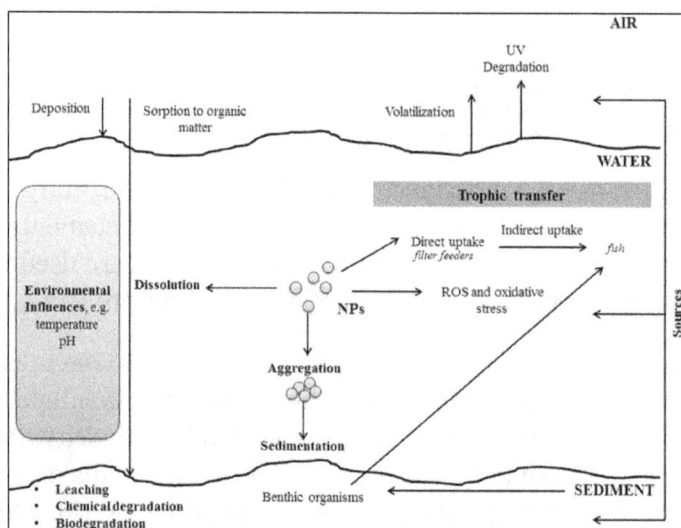

Possible pathways of nanoparticles in the aquatic environment.

Interactions between Nanoparticles and Biological Systems

The main questions scientists are currently facing are: what is the mechanism of toxic action of nanoparticles; how does the reactive surface of nanoparticles interact with "wet internal environment inside the body"; and what is the relative contribution of particle size versus particle composition in the overall toxicity of nanoparticles. Definitive answers to all these questions are currently lacking although research is underway in a number of centres. Although the chemical composition of nanomaterials is known, rearrangements, for example, of carbon into new polymeric structures, or, similarly, the restructuring of metal oxide or crystalline lattice, are worthy of toxicological considerations.

The experiences with the different forms of silica and asbestos have taught us that the physical/chemical properties of materials can be very important determinants of the toxicological potential. The very surface area and quantum chemistry effects that the nanosciences are exploring and hope to manipulate, are also known to be important in determining the manner in which biological systems function and interact with the physical world. The multitude of available in vitro studies dealing with the mechanism of nanoparticle uptake in different cell types as well as the few studies on in vivo uptake and nanoparticle distribution in animal models demonstrate that there is no single common uptake mechanism for nanoparticles. The upper size limit for the toxicity of nanoparticles (ultrafine particles) is not fully known, but is thought to lie between 65 nm and 200 nm. In vitro studies performed on cell cultures have confirmed the increased ability of nanoparticles to produce free radicals which can cause cellular damage. Generation of reactive oxygen species (ROS) upon exposure of cells to particulate matter is nowadays considered a major contributor to nanoparticle toxicity. The cell membrane, mitochondria, and cell nucleus are considered relevant for possible nanoparticle-induced toxicity.

In the light of current knowledge, it seems that the size effect is considerably more important for nanoparticle toxicity than the actual composition of the material. In contrast, in vivo pulmonary inflammation and cytotoxicity studies on rats show that TiO_2 toxicity does not depend on particle size and surface area. Over the last five years the number of papers on the ongoing work in nanotoxicology has increased exponentially. However, it is still not possible to draw any common conclusion about how nanoparticles interact with biological systems. In different in-vivo and in-vitro studies, the authors find both dose-dependent and dose-independent response to nanoparticles. The type of response is obviously related to the measured parameters. In the same study, some measured parameters may be dose-dependent, while others may not. It is interesting and worth mentioning that the highest doses do not necessarily provoke the most pronounced response, neither in in-vitro nor in in-vivo studies.

Human and Environmental Risk Assessment of Nanomaterials

Risk assessment is the evaluation of scientific information on the hazardous properties of a variety of agents, the dose-response relationship, and the extent of exposure of humans or environmental targets to these agents. The product of risk assessment is a statement about the likelihood of exposed humans and the ecosystem with all its components being harmed and to what degree (risk characterization). The following key aspects of risk assessment are addressed, as they relate to nanomaterials:

- Identification of chemical and physical properties: This is an important first step in assessing their risk. The diversity and complexity of nanomaterials makes chemical identification and characterization more difficult than with other chemicals. A broader spectrum of properties will be needed to sufficiently characterize a given nanomaterial, evaluating the hazard and assess the risk.

- Environmental fate: Fundamental properties concerning the environmental fate of nano-materials are not well understood. Models used to assess the environmental fate and exposure to conventional chemicals are not applicable to intentionally produced nanomaterials. Depending on the relevance of chemical properties or transformation, new models may have to be developed to provide estimations for new materials. However, a certain amount of reliable experimental data must be acquired before the environmental fate, transport, and multimedia partitioning of nanomaterials can be effectively modelled.

- Environmental detection and analysis: The challenge in detecting nanomaterials in the environment is compounded not only by the extremely small size of the particles, but also by their unique physical structure and physico-chemical characteristics. The variety of physical and chemical properties can significantly affect the extraction and analytical techniques that can be used for the analyses of a specific nanomaterial.

- Human and ecosystem exposure: Human and ecosystem exposure account for a series of events beginning with external mechanisms that make a chemical/nanoparticle available for absorption or other mode of entry and ending with the chemical or its metabolite reaching the target organ, depending on the nature of the chemical and route of exposure.

- Human and ecosystem effects: Assessing nanomaterial toxicity is extremely complex and multifunctional, and is potentially influenced by a variety of physico-chemical properties of nanoparticles. At the moment, there is a significant gap in our knowledge of the environmental, health and ecological implications associated with nanotechnology. However, an exponential increase in scientific papers over the last five years reflects the ongoing work and the importance of this area.

A Few known Toxic Effects of Nanomaterials

- Metal Oxide Nanoparticles: After extensive investigation, TiO_2 nanoparticles were introduced as antimicrobial particles and can be used as a coating material for medical equipment because of their antimicrobial and mechanical properties. TiO_2 can accumulate some types of oxygen, the same as hydroxyl radicals and hydrogen peroxide, which has been done by oxidation and restoration under light. The oxygen comes into contact with UV rays, then a photo catalyst produces antimicrobial features which kill all bacteria with endotoxins which, in turn, have side effects in organisms.

- The results showed that the toxicity of ZrO_2 and Al_2O_3 particles (d = 500–700 nm) increases more than TiO_2 particles (130–180 nm). Al_2O_3 coated with TiO_2 particles have shown the same toxic effects. Dendritic TiO_2 has shown higher toxicity than other forms. The toxicity of ionic metals and other chemical materials differs among cells. The larger particles tend to show greater toxicity than the smaller particles. For example, the larger TiO_2 particles cause a higher prevalence of the H-thymidine component than human monotypic macrophages. ZnO nanoparticles with a diameter of 500 to 3000 nm were placed on human fibroblasts for 24 hours. They were stained by hematoxin/eosin, which showed the extent of toxicity; the dead cells, which had been separated from the bottom of the glass, were not colored and the living cells still adhered and absorbed color. A digitizer was used to estimate the colored zone. The cells exposed to Al_2O_3, TiO_2, Fe_2O_3,

Fe_3O_4, CO_2O_3, NiO, Ga_2O_3, SnO, SnO_2, and HgO, showed no toxic effect. The results acquired from studies on ZnO, CuO, Cu_2O, Cr_2O_3, and Ag_2O show that these particles have toxic effects.

- Silver: The antimicrobial effect of silver-coated surfaces has been studied with 16 types of bacteria. Silver nanoparticles are used in medical design, especially in dentistry. For example, nanosilver crystals are used in bandages as antimicrobial agents, but the use of silver nanoparticles depends on counteracting their positive (antimicrobial effect) and negative (cellular toxicity) effects. In one study, it was observed that nanosilver (12 nm) kills E scherichia coli.

- Zinc oxide (ZnO) nanoparticles: ZnO acts as an effective UV filter when used in sun creams and textiles. Some animal studies and autoradiography have shown that ZnO nanoparticles penetrate into the skin of rats and rabbits. Particles with a diameter of 50 to 100 nm can penetrate the skin because of the intracellular space of the corneum stratum, which is about 100 nm and the distance between the two layers is 0.5 to 1 nm. During inhalation, the particles enter the deep zones of the lung where they are surrounded and excluded by macrophages before epithelial damage. The particles can attach to the epithelium (causing inflammation) and the entrance to the interstitium where they have chronic effects on cells and have the ability to move to the lymphatic nodes.

- Fullerene Toxicity: C60 was first discovered by Korto in 1985, who said that C60 has 20 dimensions in different situations and is composed of apexes and 20 faces. Some features of C60 show that it has potential for use. For example, the first use of C60 has been in optics and conductors. Also, it is used to produce various sanitary products such as creams. Research has proved the antioxidative properties of C60. In 1998, Fullerenes can be used in drug delivery systems. The preparation methods for fullerene solution are very important. To prepare this solution, fullerene must first be dissolved in polar solvent, which is able to be dissolved in water. For example, scientists use tetrahydrofuran. The quick separation of organic solvent from aqueous fullerene solution is impossible. In fact, watery solution causes toxicity. In one new study, the researchers investigated the effects of these materials on fish, and described the fullerene antioxidative effect. Cerebral damage could occur due to respiratory medicines such as D-ethyl ether. Tetrahydrofuran has an ether-like effect and is very toxic. In fact, tetrahydrofuran caused cerebral damage, but fullerene had no such effects.

- Carbon Nanotubes: Carbon nanotubes have a seamless graphite cylinder, which has featured in a number of studies, especially in medical science, it is also interesting to study because of its paradoxical effect on the body. Carbon nanotubes tend to twist in the form of a rope, which can be a problem (especially in the lungs). Nanotubes are structures that may behave as nanoparticles or fibres. For example, lung toxicity will occur at high doses of single or multiple-wall carbons, but if their amount and dose are low, inflammation will occur in the lungs. Results have shown that carbon nanotubes at high dose are toxic for organisms, and accordingly, health scientists have defined them as dangerous and suggested manipulation of the nanoparticles. One way to reduce the toxicity of carbon nanotubes is coating or functionalization. Functionalization can affect the properties of the carbon nanotubes, especially their toxicity.

- Metal based nanoparticles: (NPs) are a prominent class of NPs synthesized for their functions as semiconductors, thermoelectric and electroluminescent materials. Biomedically, these antibacterial NPs have been utilized in drug delivery systems to access areas previously inaccessible to conventional medicine. With the recent increase in interest and development of nanotechnology, many studies have been performed to assess whether the unique characteristics of these NPs, namely their small surface area to volume ratio, might negatively impact the environment upon which they were introduced. Researchers have since found that many metal and metal oxide NPs have detrimental effects on the cells with which they come into contact including but not limited to DNA breakage and oxidation, mutations, reduced cell viability, warped morphology, induced apoptosis and necrosis, and decreased proliferation. Metal Oxides such as copper oxide, uraninite, and cobalt oxide have also been found to exert significant stress on exposed DNA.

References

- Nanotechnology, technology: britannica.com, Retrieved 25 March, 2019

- Karn, Barbara; Todd Kuiken; Martha Otto (2009-12-01). "Nanotechnology and in Situ Remediation: A Review of the Benefits and Potential Risks". Environmental Health Perspectives. 117 (12): 1823–1831. doi:10.1289/ehp.0900793. ISSN 0091-6765

- Lin, C.; Ke, Y.; Chhabra, R.; Sharma, J.; Liu, Y.; Yan, H. (2011). "Synthesis and Characterization of Self-Assembled DNA Nanostructures". In Zuccheri, G. and Samorì, B (eds.). DNA Nanotechnology: Methods and Protocols. Methods in Molecular Biology. 749. pp. 1–11. doi:10.1007/978-1-61779-142-0_1. ISBN 978-1-61779-141-3. PMID 21674361

- Wagner V, Dullaart A, Bock AK, Zweck A (October 2006). "The emerging nanomedicine landscape". Nature Biotechnology. 24 (10): 1211–7. doi:10.1038/nbt1006-1211. PMID 17033654

- Nanotoxicology: amrita.edu, Retrieved 02 January, 2019

Nanomaterials

The materials whose size lies between 1 and 1000 nanometers are known as nanomaterials. Some of these are nanorods, nanoshells, nanofibers, silicene, nanosheets and nanomesh. This chapter has been carefully written to provide an easy understanding of these types of nanomaterials.

Nanomaterials are the particles (crystalline or amorphous) of organic or inorganic materials having sizes in the range of 1-100 nm. Nanomaterials are classified into nanostructured materials and nanophase/nanoparticle materials. The former refer to condensed bulk materials that are made of grains with grain sizes in the nanometer size range while the latter are usually the dispersive nanoparticles. To distinguish nanomaterials from bulk, it is vitally important to demonstrate the unique properties of nanomaterials and their prospective impacts in science and technology.

Technology in the twenty first century requires the miniaturization of devices in to nanometer sizes while their ultimate performance is dramatically enhanced. This raises many issues regarding to new materials for achieving specific functionality and selectivity. Nanotechnology is the design, fabrication and application of nanostructures or nanomaterials and the fundamental understanding of the relationships between physical properties or phenomena and material dimensions. It is a new field or a new scientific domain. Nanotechnology also promises the possibility of creating nanostructures of metastable phases with non–conventional properties including superconductivity and magnetism. Another very important aspect of nanotechnology is the miniaturization of current and new instruments, sensors and machines that will greatly impact the world we live in. Examples of possible miniaturization are computers with infinitely great power that compute algorithms to mimic human brains, biosensors that warn us at the early stage of the onset of disease and preferably at the molecular level and target specific drugs that automatically attack the diseased cells on site, nanorobots that can repair internal damage and remove chemical toxins in human bodies, nanoscaled electronics that constantly monitor our local environment.

Nanomaterials have properties that are significantly different and considerably improved relative to those of their coarser-grained counterparts. The property changes result from their small grain sizes, the large percentage of their atoms in large grain boundary environments and the interaction between the grains. Research on a variety of chemical, mechanical and physical properties is beginning to yield a glimmer of understanding of just how this interplay manifests itself in the properties of these new materials. In general, one can have nanoparticles of metals, semiconductors, dielectrics, magnetic materials, polymers or other organic compounds. Semiconductor heterostructures are usually referred to as one-dimensional artificially structured materials composed of layers of different phases/compositions. The semiconductor heterostructured material is the optimum candidate for fabricating electronic and photonic nanodevices.

It is seen that properties of these particles are quite sensitive to their sizes. This is partly connected with the fact that surface to volume ratio changes with a change in particle size. A high percentage of surface atoms introduce many size-dependent phenomena. High surface area is an important feature of nano-sized and nanoporous materials, which can be exploited in many potential industrial applications, such as separation science and catalytic processing, because of the enhanced chemical reactivity.

For optical applications, a wide range of nanostructure-based optical sources that include high performance lasers to general illumination can be fabricated. These industrial requirements can be accomplished by selecting an appropriate fabrication method of functional nanostructures with controlled size, shape and composition. However, assembling nanoparticles to form a nanostructure is a complex process. Numerous research groups are working out different synthetic strategies to find economically affordable ways for fabricating the nanostructures and simultaneously preserving the superior characteristics of the basic building units (nanoparticles) in various devices.

NANOPARTICLE

Nanoparticle is an ultrafine unit with dimensions measured in nanometres (nm; 1 nm = 10^{-9} metre). Nanoparticles exist in the natural world and are also created as a result of human activities. Because of their submicroscopic size, they have unique material characteristics, and manufactured nanoparticles may find practical applications in a variety of areas, including medicine, engineering, catalysis, and environmental remediation.

Properties of Nanoparticles

In 2008 the International Organization for Standardization (ISO) defined a nanoparticle as a discrete nano-object where all three Cartesian dimensions are less than 100 nm. The ISO standard similarly defined two-dimensional nano-objects (i.e., nanodiscs and nanoplates) and one-dimensional nano-objects (i.e., nanofibres and nanotubes). But in 2011 the Commission of the European Union endorsed a more-technical but wider-ranging definition.

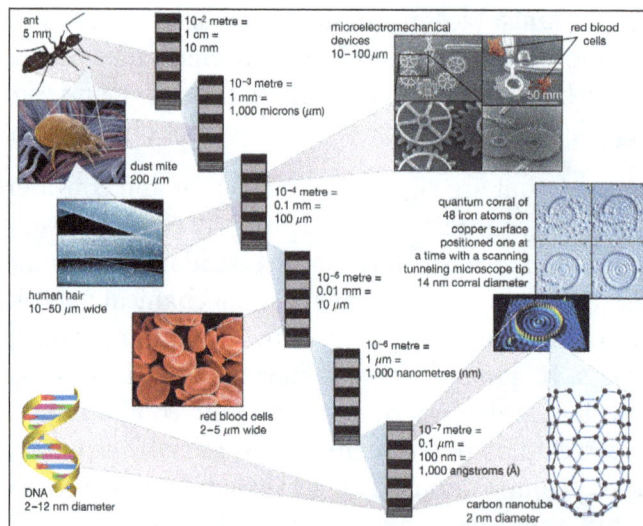

Examples from biological and mechanical realms illustrate various "orders of magnitude" (powers of 10), from 10^{-2} metre down to 10^{-7} metre.

A natural, incidental or manufactured material containing particles, in an unbound state or as an aggregate or as an agglomerate and where, for 50% or more of the particles in the number size distribution, one or more external dimensions is in the size range 1 nm–100 nm.

Under that definition a nano-object needs only one of its characteristic dimensions to be in the range 1–100 nm to be classed as a nanoparticle, even if its other dimensions are outside that range. (The lower limit of 1 nm is used because atomic bond lengths are reached at 0.1 nm).

That size range—from 1 to 100 nm—overlaps considerably with that previously assigned to the field of colloid science—from 1 to 1,000 nm—which is sometimes alternatively called the meso-scale. Thus, it is not uncommon to find literature that refers to nanoparticles and colloidal particles in equal terms. The difference is essentially semantic for particles below 100 nm in size.

Nanoparticles can be classified into any of various types, according to their size, shape, and material properties. Some classifications distinguish between organic and inorganic nanoparticles; the first group includes dendrimers, liposomes, and polymeric nanoparticles, while the latter includes fullerenes, quantum dots, and gold nanoparticles. Other classifications divide nanoparticles according to whether they are carbon-based, ceramic, semiconducting, or polymeric. In addition, nanoparticles can be classified as hard (e.g., titania [titanium dioxide], silica [silica dioxide] particles, and fullerenes) or as soft (e.g., liposomes, vesicles, and nanodroplets). The way in which nanoparticles are classified typically depends on their application, such as in diagnosis or therapy versus basic research, or may be related to the way in which they were produced.

There are three major physical properties of nanoparticles, and all are interrelated: (1) they are highly mobile in the free state (e.g., in the absence of some other additional influence, a 10-nm-diameter nanosphere of silica has a sedimentation rate under gravity of 0.01 mm/day in water); (2) they have enormous specific surface areas (e.g., a standard teaspoon, or about 6 ml, of 10-nm-diameter silica nanospheres has more surface area than a dozen doubles-sized tennis courts; 20 percent of all the atoms in each nanosphere will be located at the surface); and (3) they may exhibit what are known as quantum effects. Thus, nanoparticles have a vast range of compositions, depending on the use or the product.

Nanoparticle-based Technologies

In general, nanoparticle-based technologies centre on opportunities for improving the efficiency, sustainability, and speed of already-existing processes. That is possible because, relative to the materials used traditionally for industrial processes (e.g., industrial catalysis), nanoparticle-based technologies use less material, a large proportion of which is already in a more "reactive" state. Other opportunities for nanoparticle-based technologies include the use of nanoscale zero-valent iron (NZVI) particles as a field-deployable means of remediating organochlorine compounds, such as polychlorinated biphenyls (PCBs), in the environment. NZVI particles are able to permeate into rock layers in the ground and thus can neutralize the reactivity of organochlorines in deep aquifers. Other applications of nanoparticles are those that stem from manipulating or arranging matter at the nanoscale to provide better coatings, composites, or additives and those that exploit the particles' quantum effects (e.g., quantum dots for imaging, nanowires for molecular electronics, and technologies for spintronics and molecular magnets).

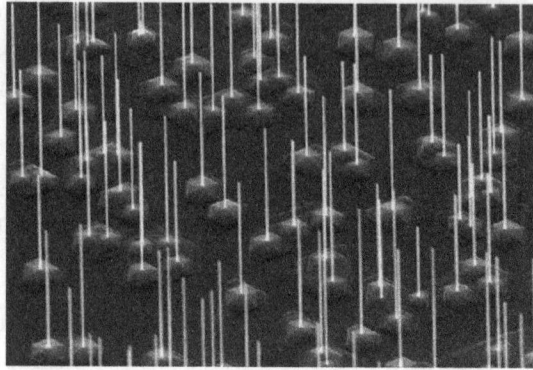

Nanowires as seen by a field-emission microscope.

Nanoparticle Applications in Materials

Many properties unique to nanoparticles are related specifically to the particles' size. It is therefore natural that efforts have been made to capture some of those properties by incorporating nanoparticles into composite materials. An example of how the unique properties of nanoparticles have been put to use in a nanocomposite material is the modern rubber tire, which typically is a composite of a rubber (an elastomer) and an inorganic filler (a reinforcing particle), such as carbon black or silica nanoparticles.

For most nanocomposite materials, the process of incorporating nanoparticles is not straightforward. Nanoparticles are notoriously prone to agglomeration, resulting in the formation of large clumps that are difficult to redisperse. In addition, nanoparticles do not always retain their unique size-related properties when they are incorporated into a composite material.

Despite the difficulties with manufacture, the use of nanomaterials grew markedly in the early 21st century, with especially rapid growth in the use of nanocomposites. Nanocomposites were employed in the development and design of new materials, serving, for example, as the building blocks for new dielectric (insulating) and magnetic materials.

Polymers

Similar to the way in which carbon and silica nanoparticles have been used as fillers in rubber to improve the mechanical properties of tires, such particles and others, including nanoclays, have been incorporated into polymers to improve their strength and impact resistance. In the early 21st century, increasing use of non-petroleum-based polymers that were derived from natural sources drove the development of "all-natural" nanocomposite polymers. Such materials incorporate a biopolymer derived from an alginate (a carbohydrate found in the cell wall of brown algae), cellulose, or starch; the biopolymer is used in conjunction with a natural nanoclay or a filler derived from the shells of crustaceans. The materials are biodegradable and do not leave behind potentially harmful or nonnatural residues.

Food Packaging

Nanoparticles have been increasingly incorporated into food packaging to control the ambient atmosphere around food, keeping it fresh and safe from microbial contamination. Such composites use nanoflakes of clays and claylike particles, which slow down the ingress of moisture and reduce gas transport across the packaging film. It is also possible to incorporate nanoparticles with apparent antimicrobial effects (e.g., nanocopper or nanosilver) into such packaging. Nanoparticles that exhibit antimicrobial activity had also been incorporated into paints and coatings, making those products particularly useful for surfaces in hospitals and other medical facilities and in areas of food preparation.

Flame Retardants

Nanoparticles were explored for their potential to replace additives based on flammable organic halogens and phosphorus in plastics and textiles. Studies had suggested that, in the event of a serious fire, products with nanoclays and hydroxide nanoparticles were associated with fewer emissions of harmful fumes than products containing certain other types of additives.

Batteries and Supercapacitors

The ability to engineer nanocomposite materials to have very high internal surface areas for storing electrical charge in the form of small ions or electrons has made them especially valuable for use in batteries and supercapacitors. Indeed, nanocomposite materials have been synthesized for various applications involving electrodes. Composite materials based on carbon nanotubes and layered-type materials, such as graphene, were also researched extensively, making their first appearances in commercial devices in the early 2000s.

Nanoceramics

A long-term objective in materials science had been to transform ceramics that are brittle and prone to cracking into tougher, more resilient materials. By the early 21st century, researchers had achieved that goal by incorporating an effective blend of nanoparticles into ceramics materials. Other new ceramics materials that were under development included all-ceramic or polymer-ceramic blends, which combined the unique functional (e.g., electrical, magnetic, or mechanical) properties of a nanocomposite material with the properties of ceramics materials.

Light Control

In the 1990s the development of blue light-emitting diodes (LEDs), which had the potential to produce white light at significantly reduced costs, inspired a revolution in lighting. Blue LEDs brought about a need for composite materials that could be used to coat the diodes to convert blue light into other wavelengths (such as red, yellow, or green) in order to achieve white light. One way of obtaining the desired light is by leveraging the size or quantum effect of small semiconducting particles. The application of such particles facilitated the development of nanocomposite polymers for greenhouse enclosures; the polymers optimize plant growth by effectively converting wavelengths of full-spectrum sunlight into the red and blue wavelengths used in photosynthesis. Light conversion in the above cases is achieved with submicron particles of inorganic phosphor materials incorporated into the polymer.

Nanoparticle Applications in Medicine

The small size of nanoparticles is especially advantageous in medicine; nanoparticles can not only circulate widely throughout the body but also enter cells or be designed to bind to specific cells. Those properties have enabled new ways of enhancing images of organs as well as tumours and other diseased tissues in the body. They also have facilitated the development of new methods of delivering therapy, such as by providing local heating (hyperthermia), by blocking vasculature to diseased tissues and tumours, or by carrying payloads of drugs.

Magnetic nanoparticles have been used to replace radioactive technetium for tracking the spread of cancer along lymph nodes. The nanoparticles work by exploiting the change in contrast brought about by tiny particles of superparamagnetic iron oxide in magnetic resonance imaging (MRI). Such particles also can be used to kill tumours via hyperthermia, in which an alternating magnetic field causes them to heat and destroy tissue on a local scale.

Nanoparticles can be designed to enhance fluorescent imaging or to enhance images from positron emission tomography (PET) or ultrasound. Those methods typically require that the nanoparticle be able to recognize a particular cell or disease state. In theory, the same idea of targeting could be used in aiding the precise delivery of a drug to a given disease site. The drug could be carried via a nanocapsule or a liposome, or it could be carried in a porous nanosponge structure and then held by bonds at the targeted site, thereby allowing the slow release of drug. The development of nanoparticles to aid in the delivery of a drug to the brain via inhalation holds considerable promise for the treatment of neurological disorders such as Parkinson disease, Alzheimer disease, and multiple sclerosis.

Nanoparticles and nanofibres play an important part in the design and manufacture of novel scaffold structures for tissue and bone repair. The nanomaterials used in such scaffolds are biocompatible. For example, nanoparticles of calcium hydroxyapatite, a natural component of bone, used in combination with collagen or collagen substitutes could be used in future tissue-repair therapies.

Nanoparticles also have been used in the development of health-related products. For example, a sunscreen known as Optisol, invented at the University of Oxford in the 1990s, was designed with the objective of developing a safe sunscreen that was transparent in visible light but retained ultraviolet-blocking action on the skin. The ingredients traditionally used in sunscreens were based on large particles of either zinc oxide or titanium dioxide or contained an organic sunlight-absorbing compound. However, those materials were not satisfactory: zinc oxide and titanium dioxide are very potent photocatalysts, and in the presence of water and sunlight they generate free radicals, which have the potential to damage skin cells and DNA (deoxyribonucleic acid). Scientists proceeded to develop a nanoparticle form of titanium oxide that contained a small amount of manganese. Studies indicated that the nanoparticle-based sunscreen was safer than sunscreen products manufactured by using traditional materials. The improvement in safety was attributed to the introduction of manganese, which changed the semiconducting properties of the compound from n-type to p-type, thus shifting its Fermi level, or oxidation-reduction properties, and making the generation of free radicals less likely.

Treatments and diagnostic approaches based on the use of nanoparticles are expected to have important benefits for medicine in the future, but the use of nanoparticles also presents significant challenges, particularly regarding impacts on human health. For example, little is known about the fate of nanoparticles that are introduced into the body or whether they have undesirable effects on the body. Extensive clinical trials are needed in order to fully address concerns about the safety and effectiveness of nanoparticles used in medicine. There also are manufacturing problems to be overcome, such as the ability to produce nanoparticles under sterile conditions, which is required for medical applications.

Manufacture of Nanoparticles

Nanoparticles are made by one of three routes: by comminution (the pulverization of materials), such as through industrial milling or natural weathering; by pyrolysis (incineration); or by sol-gel synthesis (the generation of inorganic materials from a colloidal suspension). Comminution is known as a top-down approach, whereas the sol-gel process is a bottom-up approach. Examples of those three processes (comminution, pyrolysis, and sol-gel synthesis) include the production of titania nanoparticles for sunscreens from the minerals anatase and rutile, the production of fullerenes or fumed silica and the production of synthetic (or Stöber) silica, of other "engineered" oxide nanoparticles, and of quantum dots. For the generation of small nanoparticles, comminution is a very inefficient process.

Detection, Characterization and Isolation

The detection and characterization of nanoparticles present scientists with particular challenges. Being of a size that is at least four to seven times smaller than the wavelength of light means that individual nanoparticles cannot be detected by the human eye, and they are observable under optical microscopes only in liquid samples under certain conditions. Thus, in general, specialized techniques are required to see them, and none of those approaches is currently field-deployable.

Techniques to detect and characterize nanoparticles fall into two categories: direct, or "real space," and indirect, or "reciprocal space." Direct techniques include transmission electron microscopy (TEM), scanning electron microscopy (SEM), and atomic force microscopy (AFM). Those techniques can image nanoparticles, directly measure sizes, and infer shape information, but they are limited to studying only a few particles at a time. There are also significant issues surrounding sample preparation for electron microscopy. In general, however, those techniques can be quite effective for obtaining basic information about a nanoparticle.

Indirect techniques use X-rays or neutron beams and obtain their information by mathematically analyzing the radiation scattered or diffracted by the nanoparticles. The techniques of greatest relevance to nanoscience are small-angle X-ray scattering (SAXS) and small-angle neutron scattering (SANS), along with their surface-specific analogues GISAXS and GISANS, where GI is "grazing incidence," and X-ray or neutron reflectometry (XR/NR). The advantage of those techniques is that they are able to simultaneously sample and average very large numbers of nanoparticles and often do not require any particular sample preparation. Indirect techniques have many applications. For example, in studies of nanoparticles in raw sewage, scientists used SANS measurements, in which neutrons readily penetrated the turbid sewage and scattered strongly from the nanoparticles, to follow the aggregation behavior of the particles over time.

The isolation of nanoparticles from colloidal and larger matter involves specialized techniques, such as ultra centrifugation and field-flow fractionation. Such laboratory-based techniques are normally coupled to standard spectroscopic instrumentation to enable particular types of chemical characterization.

Nanoparticles in the Environment

Nanoparticles occur naturally in the environment in large volumes. For example, the sea emits an aerosol of salt that ends up floating around in the atmosphere in a range of sizes, from a few nanometres upward, and smoke from volcanoes and fires contains a huge variety of nanoparticles, many of which could be classified as dangerous to human health. Dust from deserts, fields, and so on also has a range of sizes and types of particles, and even trees emit nanoparticles of hydrocarbon compounds such as terpenes (which produce the familiar blue haze seen in forests, from which the Great Smoky Mountains in the United States get their name).

Human-made (anthropogenic) nanoparticles are emitted by large industrial processes, and in modern life it is particles from power stations and from jet aircraft and other vehicles (namely, those powered by internal-combustion engines; car tires are also a factor) that constitute the major fraction of nanoparticle emissions. Types of nanoparticles that are emitted include partially burned hydrocarbons (in soot), ceria (cerium oxide; from vehicle exhaust catalysts), metallic dust (from brake linings), calcium carbonate (in engine lubricating oils), and silica (from car tires). Other sources of nanoparticles to the environment include the semiconductor industry, domestic and industrial wastewater discharges, the health care industry, and the photographic industry. However, all those emission levels are still considered to be lower than the levels of nanoparticles produced through natural processes. Indeed, recent human-made particles contribute only a small amount to air and water pollution.

Understanding the relationship between nanoparticles and the environment forms an important area of research. There are several mechanisms by which nanoparticles are believed to affect the environment negatively. Two scenarios that are under investigation are the possibilities (1) that the mobility and sorptive capacity of nanoparticles (natural or human-made) make them potent vectors (carriers) in the transport of chemical pollutants (e.g., phosphorus from sewage and agriculture), particularly in rivers and lakes, and (2) that some nanoparticles are able to reduce the functioning of (and may even disrupt or kill) naturally occurring microbial communities, as well as microbial communities that are employed in industrial processes (e.g., those that are used in sanitation processes, including sewage treatment).

Nanoparticles also can have beneficial impacts on the environment and appear to contribute to natural processes. Thus, in addition to the potential use of nanoparticles to remove chemical contaminants from the environment, scientists are investigating how nanoparticles interact with all life-forms—from fungi to microbes, algae, plants, and higher-order animals. That type of study is essential not only to improving scientists' knowledge of nanoparticles but also to gaining a more complete understanding of life on Earth, since the soil is naturally full of nanoparticles, in a richly diverse environment.

Health Effects of Nanoparticles

Humans have evolved to cope with most naturally occurring nanoparticles. However, some nanoparticles, generated as a result of certain human activities such as tobacco smoking and fires, account for many premature deaths as a result of lung damage. For example, fires from the types of cooking stoves used in developing countries are known to emit fine particles and lead to early mortality, especially among women who routinely work near the stoves.

Laboratory and clinical investigations of the effects of nanoparticles on health have been somewhat controversial and remain largely inconclusive. Most studies in animals have involved nanoparticle inhalation, and the dosages have been very large. The results of those studies have indicated that large quantities of nanoparticles can cause cellular damage in the lungs, with lung cells absorbing the particles and becoming damaged or undergoing genetic mutation. Animal studies involving the ingestion of nanoparticles in food or water suggest that nanoparticles can also affect health in other ways. For example, consumption of the food additive E171, which consists of titanium dioxide nanoparticles, is associated with changes in gut microbiota (bacteria occurring in the gut), potentially contributing to the development of conditions such as inflammatory bowel disease.

In humans, the health effects of typical exposure levels—those that are encountered by most persons during daily activities—remain unknown. Nonetheless, there is a general awareness of the problems that might occur upon excess exposure to nanoparticles, and, thus, most manufacturers of such particles take serious precautions to avoid exposure of their workers. Efforts have been made to educate the public in the use of nanoparticle-containing products. The existence of pressure groups has also helped to ensure nanoparticle safety compliance among manufacturers. However, nanoparticles offer tremendous potential for new or improved forms of health care treatment. That has spawned a new field of science called nanomedicine.

Nanorod

In nanotechnology, nanorods are one morphology of nanoscale objects. Each of their dimensions range from 1–100 nm. They may be synthesized from metals or semiconducting materials. Standard aspect ratios (length divided by width) are 3-5. Nanorods are produced by direct chemical synthesis. A combination of ligands act as shape control agents and bond to different facets of the nanorod with different strengths. This allows different faces of the nanorod to grow at different rates, producing an elongated object.

One potential application of nanorods is in display technologies, because the reflectivity of the rods can be changed by changing their orientation with an applied electric field. Another application is for microelectromechanical systems (MEMS). Nanorods, along with other noble metal nanoparticles, also function as theragnostic agents. Nanorods absorb in the near IR, and generate heat when excited with IR light. This property has led to the use of nanorods as cancer therapeutics. Nanorods can be conjugated with tumor targeting motifs and ingested. When a patient is exposed to IR light (which passes through body tissue), nanorods selectively taken up by tumor cells are locally heated, destroying only the cancerous tissue while leaving healthy cells intact.

Nanorods based on semiconducting materials have also been investigated for application as energy harvesting and light emitting devices. In 2006, Ramanathan et al. demonstrated electric-field mediated tunable photoluminescence from ZnO nanorods, with potential for application as novel sources of near-ultraviolet radiation.

Synthesis

ZnO Nanorods

Zinc oxide (ZnO) nanorod, also known as nanowire, has a direct bandgap energy of 3.37 eV, which is similar to that of GaN, and it has an excitation binding energy of 60 meV. The optical bandgap of ZnO nanorod can be tuned by changing the morphology, composition, size etc. Recent years, ZnO nanorods have been intensely used to fabricate nano-scale electronic devices, including field effect transistor, ultraviolet photodetector, Schottky diode, and ultra-bright light-emitting diode (LED). Various methods have been developed to fabricate the single crystalline, wurtzite ZnO nanorods. Among those methods, growing from vapor phase is the most developed approach. In a typical growth process, ZnO vapor is condensed onto a solid substrate. ZnO vapor can be generated by three methods: thermal evaporation, chemical reduction, and Vapor-Liquid-Solid (VLS) method. In the thermal evaporation method, commercial ZnO powder is mixed with SnO_2 and evaporated by heating the mixture at elevated temperature. In the chemical reduction method, zinc vapor, generated by the reduction of ZnO, is transferred to the growth zone, followed by reoxidation to ZnO. The VLS process, originally proposed in 1964, is the most commonly used process to synthesize single crystalline ZnO nanorods. In a typical process, catalytic droplets are deposited on the substrate and the gas mixtures, including Zn vapor and a mixture of CO/CO_2, react at the catalyst-substrate interface, followed by nucleation and growth. Typical metal catalysts involve gold, copper, nickel, and tin. ZnO nanowires are grown epitaxially on the substrate and assemble into monolayer arrays. Metal-organic chemical vapor deposition (MOCVD) has also been recently developed. No catalyst is involved in this process and the growth temperature is at 400 ~500 °C, i.e. considerably milder conditions compared to the traditional vapor growth method. Moreover, metal oxide nanorods (ZnO, CuO, Fe_2O_3, V_2O_5, others) can be simply made by heating initial metal in air in a Thermal oxidation process. For example, to make a dense "carpet" of CuO nanorods it was found to be enough to heat Cu foil in air at 420 °C.

Gold Nanorods

The seed-mediated growth method is the most common and achieved method for synthesizing high-quality gold nanorods. A typical growth protocol involves the addition of citrate-capped gold nanospheres, served as seeds, to the bulk $HAuCl_4$ growth solution. The growth solution is obtained by the reduction of $HAuCl_4$ with ascorbic acid in the presence of cetyltrimethylammonium bromide (CTAB) surfactant and silver ions. Longer nanorods (up to an aspect ratio of 25) can be obtained in the absence of silver nitrate by use of a three-step addition procedure. In this protocol, seeds are sequentially added to growth solution in order to control the rate of heterogeneous deposition and thereby the rate of crystal growth.

The shortcoming of this method is the formation of gold nanospheres, which requires non-trivial separations and cleanings. In one modifications of this method sodium citrate is replaced with a stronger CTAB stabilizer in the nucleation and growth procedures. Another improvement is to introduce silver ions to the growth solution, which results in the nanorods of aspect ratios less than five in greater than 90% yield. Silver, of a lower reduction potential than gold, can be reduced on the surface of the rods to form a monolayer by underpotential deposition. Here, silver deposition competes with that of gold, thereby retarding the growth rate of specific crystal facets, allowing for one-directional growth and rod formation. Another shortcoming of this method is the high toxicity of CTAB. Polymers, such as Polyethylene glycol (PEG), Polyallylamine hydrochloride (PAH) coating, or dietary fibers, such as chitosan, to displace the CTAB out from the nanorod surface without affecting the stability has been reported.

Cation Exchange

Cation exchange is a conventional but promising technique for new nanorod synthesis. Cation exchange transformations in nanorods are kinetically favorable and often shape-conserving. Compared to bulk crystal systems, the cation exchange of nanorods is million-times faster due to high surface area. Existing nanorods serve as templates to make a variety of nanorods that are not accessible in traditional wet-chemical synthesis. Furthermore, complexity can be added by partial transformation, making nanorod heterostructures.

Nanoshell

A nanoshell, or rather a nanoshell plasmon, is a type of spherical nanoparticle consisting of a dielectric core which is covered by a thin metallic shell (usually gold). These nanoshells involve a quasiparticle called a plasmon which is a collective excitation or quantum plasma oscillation where the electrons simultaneously oscillate with respect to all the ions.

The simultaneous oscillation can be called plasmon hybridization where the tunability of the oscillation is associated with mixture of the inner and outer shell where they hybridize to give a lower energy or higher energy. This lower energy couples strongly to incident light, whereas the higher energy is an anti-bonding and weakly combines to incident light. The hybridization interaction is stronger for thinner shell layers, hence, the thickness of the shell and overall particle radius determines which wavelength of light it couples with. Nanoshells can be varied across a broad range of the light spectrum that spans the visible and near infrared regions. The interaction of light and nanoparticles affects the placement of charges which affects the coupling strength. Incident light polarized parallel to the substrate gives a s-polarization, hence the charges are further from the substrate surface which gives a stronger interaction between the shell and core. Otherwise, a p-polarization is formed which gives a more strongly shifted plasmon energy causing a weaker interaction and coupling.

Production

A state of the art method for synthesizing gold nanoshells is the use of the Microfluidic Composite Foams. This method has the potential to replace the standard lithographic method of synthesizing plasmonic nanoshells. The production process described below was an experiment performed by Suhanya Duraiswamy and Saif A. Khan of the Department of Chemical and Biomolecular Engineering in Singapore. Although this method was an experiment, it represents the future of nanoshells synthesis.

The materials required for the production of the nanoshells are the following; Tetraethyl orthosilicate, ammonium hydroxide, hydroxylamine hydrochloride, 3-aminopropyl tris, hydrogentetra-chloroaurate(III) trihydrate, tetrakis(hydroxymethyl) phosphonium chloride, sodium hydroxide, potassium carbonate, ethanol, Ultrapure water and glassware washed in aqua regia and rinsed thoroughly in water.

The first step in synthesizing nanoshells in this method is by creating the device for the reaction to take place within. Microfluidic device patterns were fabricated onto silicon wafers by standard photolithography using negative photoresist SU-8 2050. Devices were subsequently molded in poly(dimethyl siloxane) (PDMS) using the soft lithography technique. Briefly, PDMS was molded onto the SU-8 masters at 70 °C for 4 h, peeled, cut, and cleaned. Inlet and outlet holes (1/16-in. o.d.) were punched into the device. The microchannels were irreversibly bonded to a glass slide precoated with a thin layer of PDMS after a brief 35 s air plasma treatment. The microchannels have rectangular cross-section and are 300 μm wide, 155 μm deep, and 0.45 m long.

The actual production of the nanoparticles involves pumping "silicone oil, a mixture of gold-seeded silica particles and gold-plating solution and reducing agent solution to the microfluidic device while nitrogen gas was delivered from a cylinder." The plating solution was then left to age, in a controlled environment, for longer than 24 hours. After the aging process, the fluid is collected from the Microfluidic Device and placed in a centrifuge. The resulting liquid has a layer of oil on the surface with a solution below that contains the nanoshells.

The reason this method is revolutionary is that the size and relative thickness of the gold nanoshell can be controlled by changing the amount of time the reaction is allowed to take place as well as the concentration of the plating solution. Thereby allowing researchers to tailor the particles to suit their given needs. Albeit for optics or cancer treatment.

Cancer Treatment

Gold-shelled nanoparticles, which are spherical nanoparticles with silica and liposome cores and gold shells, are used in cancer therapy and bio-imaging enhancement. Theranostic probes – capable of detection and treatment of cancer in a single treatment - are nanoparticles that have binding sites on their shell that allow them to attach to a desired location (typically cancerous cells) then can be imaged through dual modality imagery (an imaging strategy that uses x-rays and radionuclide imaging) and through near-infrared fluorescence. The reason gold nanoparticles are used is due to their vivid optical properties which are controlled by their size, geometry, and their surface plasmons. Gold nanoparticles (such as AuNPs) have the benefit of being biocompatible and the flexibility to have multiple different molecules, and fundamental materials, attached to their shell (almost anything that can normally be attached to gold can be attached to the gold nano-shell, which can be used in helping identifying and treating cancer). The treatment of cancer is possible only because of the scattering and absorption that occurs for plasmonics. Under scattering, the gold-plated nano-particles become visible to imaging processes that are tuned to the correct wavelength which is dependent upon the size and geometry of the particles. Under absorption, photothermal ablation occurs, which heats the nanoparticles and their immediate surroundings to temperatures capable of killing the cancer cells. This is accomplished with minimal damage to cells in the body due to the utilization of the "water window" (the spectral range between 800 and 1300 nm). As the human body is mostly water, this optimizes the light used versus the effects rendered.

These gold nanoshells are shuttled into tumors by the use of phagocytosis, where phagocytes engulf the nanoshells through the cell membrane to form an internal phagosome, or macrophage. After this it is shuttled into a cell and enzymes are usually used to metabolize it and shuttle it back out of the cell. These nanoshells are not metabolized so for them to be effective they just need to be within the tumor cells and photo-induced cell death is used to terminate the tumor cells.

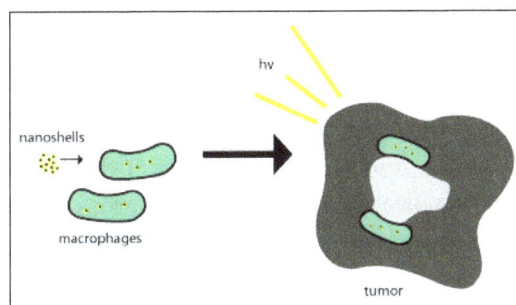

Nanoparticle-based therapeutics have been successfully delivered into tumors by exploiting the enhanced permeability and retention effect, a property that permits nanoscale structures to be taken up passively into tumors without the assistance of antibodies. Delivery of nanoshells into the important regions of tumors can be very difficult. This is where most nanoshells try to exploit the tumor's natural recruitment of monocytes for delivery as seen in the above figure. This delivery system is called a "Trojan Horse".

This process works so well since tumors are about ¾ macrophages and once monocytes are brought into the tumor, it differentiates into macrophages which would also be need to maintain the cargo nanoparticles. Once the nanoshells are at the necrotic center, near-infrared illumination is used to destroy the tumor associated macrophages.

Additionally, these nanoparticles can be made to release antisense DNA oligonucleotides when under photo-activation. These oligonucleotides are used in conjunction with the photo-thermal ablation treatments to perform gene-therapy. This is accomplished because nanoparticle complexes are delivered inside of cells then undergo light induced release of DNA from their surface. This will allow for the internal manipulation of a cell and provide a means for monitoring a group cells return to equilibrium.

Another example of nanoshell plasmonics in cancer treatment involves placing drugs inside of the nanoparticle and using it as a vehicle to deliver toxic drugs to cancerous sites only. This is accomplished by coating the outside of a nanoparticle with iron oxide (allowing for easy tracking with an MRI machine), then once the area of the tumor is coated with the drug-filled nanoparticles, the nanoparticles can be activated using resonant light waves to release the drug.

Plasmonic Nanoparticles

Plasmonic nanoparticles are particles whose electron density can couple with electromagnetic radiation of wavelengths that are far larger than the particle due to the nature of the dielectric-metal interface between the medium and the particles: unlike in a pure metal where there is a maximum limit on what size wavelength can be effectively coupled based on the material size.

What differentiates these particles from normal surface plasmons is that plasmonic nanoparticles also exhibit interesting scattering, absorbance, and coupling properties based on their geometries and relative positions. These unique properties have made them a focus of research in many applications including solar cells, spectroscopy, signal enhancement for imaging, and cancer treatment. As well owing to their high sensitivity appear to be good candidates for designing mechano-optical instrumentation.

Plasmons are the oscillations of free electrons that are the consequence of the formation of a dipole in the material due to electromagnetic waves. The electrons migrate in the material to restore its initial state; however, the light waves oscillate, leading to a constant shift in the dipole that forces the electrons to oscillate at the same frequency as the light. This coupling only occurs when the frequency of the light is equal to or less than the plasma frequency and is greatest at the plasma frequency that is therefore called the resonant frequency. The scattering and absorbance cross-sections describe the intensity of a given frequency to be scattered or absorbed. Many fabrication processes or chemical synthesis methods exist for preparation of such nanoparticles, depending on the desired size and geometry.

The nanoparticles can form clusters (the so-called "plasmonic molecules") and interact with each other to form cluster states. The symmetry of the nanoparticles and the distribution of the electrons within them can affect a type of bonding or antibonding character between the nanoparticles similarly to molecular orbitals. Since light couples with the electrons, polarized light can be used to control the distribution of the electrons and alter the mulliken term symbol for the irreducible representation. Changing the geometry of the nanoparticles can be used to manipulate the optical activity and properties of the system, but so can the polarized light by lowering the symmetry of the conductive electrons inside the particles and changing the dipole moment of the cluster. These clusters can be used to manipulate light on the nano scale.

Theory

The quasistatic equations that describe the scattering and absorbance cross-sections for very small spherical nanoparticles are:

$$\sigma_{scatt} = \frac{8\pi}{3} k^4 R^6 \left| \frac{\varepsilon_{particle} - \varepsilon_{medium}}{\varepsilon_{particle} + 2\varepsilon_{medium}} \right|^2$$

$$\sigma_{abs} = 4\pi k R^3 \, \text{Im} \left| \frac{\varepsilon_{particle} - \varepsilon_{medium}}{\varepsilon_{particle} + 2\varepsilon_{medium}} \right|$$

where, k is the wavenumber of the electric field, R is the radius of the particle, ε_{medium} is the relative permittivity of the dielectric medium and $\varepsilon_{particle}$ is the relative permittivity of the nanoparticle defined by,

$$\varepsilon_{particle} = 1 - \frac{\omega_p^2}{\omega^2 + i\omega\gamma}$$

also known as the Drude Model for free electrons where ω_p is the plasma frequency, γ is the relaxation frequency of the charge carries, and ω is the frequency of the electromagnetic radiation. This equation is the result of solving the differential equation for a harmonic oscillator with a driving force proportional to the electric field that the particle is subjected to.

It logically follows that the resonance conditions for these equations is reached when the denominator is around zero such that,

$$\varepsilon_{particle} + 2\varepsilon_{medium} \approx 0$$

When this condition is fulfilled the cross-sections are at their maximum.

These cross-sections are for single, spherical particles. The equations change when particles are non-spherical, or are coupled to 1 or more other nanoparticles, such as when their geometry changes. This principle is important for several applications.

Rigorous electrodynamic analysis of plasma oscillations in a spherical metal nanoparticle of a finite size was performed in.

Applications

Plasmonic Solar Cells

Due to their ability to scatter light back into the photovoltaic structure and low absorption, plasmonic nanoparticles are under investigation as a method for increasing solar cell efficiency. Forcing more light to be absorbed by the dielectric increases efficiency.

Plasmons can be excited by optical radiation and induce an electric current from hot electrons in materials fabricated from gold particles and light-sensitive molecules of porphin, of precise sizes and specific patterns. The wavelength to which the plasmon responds is a function of the size and spacing of the particles. The material is fabricated using ferroelectric nanolithography. Compared to conventional photoexcitation, the material produced three to 10 times the current.

Spectroscopy

In the past 5 years plasmonic nanoparticles have been explored as a method for high resolution spectroscopy. One group utilized 40 nm gold nanoparticles that had been functionalized such that they would bind specifically to epidermal growth factor receptors to determine the density of those receptors on a cell. This technique relies on the fact that the effective geometry of the particles change when they appear within one particle diameter (40 nm) of each other. Within that range, quantitative information on the EGFR density in the cell membrane can be retrieved based on the shift in resonant frequency of the plasmonic particles.

Cancer Treatment

Preliminary research indicates that the absorption of gold nanorods functionalized with epidermal growth factor is enough to amplify the effects of low power laser light such that it can be used for targeted radiation treatments.

MAGNETOELASTIC FILAMENTS

Magnetoelastic filaments are one-dimensional composite structures that exhibit both magnetic and elastic properties. Interest in these materials tends to focus on the ability to precisely control mechanical events using an external magnetic field. Like piezoelectricity materials, they can be used as actuators, but do not need to be physically connected to a power source. The conformations adopted by magnetoelastic filaments are dictated by the competition between its elastic and magnetic properties.

Mechanical Behavior

Magnetic Nanochains

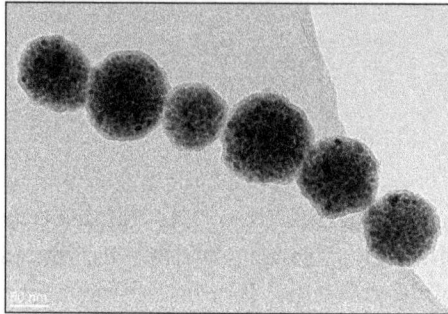

TEM image of well-defined magnetic nanochain
composed of six magnetic nanobeads.

Magnetic nanochains are a new class of magnetoresponsive and superparamagnetic nanostructures with highly anisotropic shapes which can be manipulated using magnetic field and magnetic field gradient. Such nanochains consist of self-assembled nanoparticle clusters which are magnetically assembled and fixated into a chain. Among the various linking methods used are silica coating, polyacrylic acid (PAA) coating, tetraethoxysilane condensation, biotinylation or glucose decomposition. Typically, the primary building blocks of these nanostructures are individual superparamagnetic iron oxide nanoparticles (SPIONs). Nanoparticle clusters which are composed of a number of individual magnetic nanoparticles (ca. 100 SPIONs) are known as magnetic nanobeads with a diameter of 50–200 nanometers.

The force exerted on a particle depends on the strength, direction, and dynamics of the applied magnetic field as well as the position and orientation of local magnetic dipoles. Dynamic magnetic fields allow for the greatest range of control over chain shape. Of principal interest is the force exerted on the ends of the chain as a result of a dynamic field. The effect of Larmor precession with a row of magnetic colloids results in dynamic interactions dependent on the field precession angle. In fact, sweeping through the magic angle flips sign of the dipole-dipole interaction. In a field precessing quickly around the z-axis, the force exerted on the end of the chain is given by,

$$F_{end} = \frac{3\mu^2}{\sigma^4}\sin(2(\omega - \frac{d\psi}{dt})t)\mathbf{N} - \gamma\frac{d\psi}{dt}\mathbf{N}$$

where, μ is the dipole moment, σ is the bead diameter, ω is the angular frequency of the field precession, $\frac{d\psi}{dt}$ is the rate of change of the filament path, γ is the viscous drag coefficient and $\mathbf{N}$ is the unit vector of the plane perpendicular to the tangent of the filament curve. This produces a periodic magnetic force. However, under fast precession, the second term remains non-zero and scales with ω^{-1}. At low ω, the magnetic torque dominates and the chain winds around itself. With a high ω, the bending modulus dominates the energetic landscape and filaments form branched gels with a field-dependent bulk modulus.

The applied load on a filament is generally limited by the polymer linking method. The elastic strain regime for a simple covalently linked filament is short and are taken as inextensible under most conditions. If tensile forces become too large, plastic deformation can occur usually resulting in bond breaking and polymer disentanglement. These irreversible changes can result in the permanent change in the bending modulus which ultimately effects the filament performance.

Alloy Nanopillars

Using etching techniques such as focused ion beam milling, micro- or nano-sized pillars can be formed in magnetic materials. However, repeated bending of crystal pillars can cause defect formation and fatigue damage. This damage comes from the nucleation of cracks on the pillars surface, even in the elastic regime, due to localized plasticity. Crack propagation during successive compression and tension cycles can lead to pillar fracture. This is similar to what can be seen in cantilever magnetometry when operating under strong fields. Because of this, it is desirable to link smaller magnetic particles together with tougher, elastic materials, such as a polymer, rather than use a continuous alloy filament.

Applications

The fabrication of magnetic nanochains with controlled aspect ratio, a uniform size, and a well-defined shape is the focus of many world-leading research groups and high-tech companies. The magnetic nanochains possess attractive properties which are significant added value for many potential uses including magneto-mechanical actuation-associated nanomedicines in low and super-low frequency alternating magnetic field. Such structures are used in a variety of applications, such as imaging and drug delivery. Other applications are shown below:

- Mechanical sensors for testing the elastic moduli of biomolecules and nanostructures,

- Microactuation,

- MRI imaging,

- Drug delivery,

- Responsive coatings.

NANOFIBER

Nanofibers are fibers with diameters in the nanometer range. Nanofibers can be generated from different polymers and hence have different physical properties and application potentials. Examples of natural polymers include collagen, cellulose, silk fibroin, keratin, gelatin and polysaccharides such as chitosan and alginate. Examples of synthetic polymers include poly(lactic acid) (PLA), polycaprolactone (PCL), polyurethane (PU), poly(lactic-co-glycolic acid) (PLGA), poly(3-hydroxybutyrate-co-3-hydroxyvalerate) (PHBV), and poly(ethylene-co-vinylacetate) (PEVA). Polymer chains are connected via covalent bonds. The diameters of nanofibers depend on the type of polymer used and the method of production. All polymer nanofibers are unique for their large surface area-to-volume ratio, high porosity, appreciable mechanical strength, and flexibility in functionalization compared to their microfiber counterparts.

There exist many different methods to make nanofibers, including drawing, electrospinning, self-assembly, template synthesis, and thermal-induced phase separation. Electrospinning is the most commonly used method to generate nanofibers because of the straightforward setup, the ability to mass-produce continuous nanofibers from various polymers, and the capability to generate ultrathin fibers with controllable diameters, compositions, and orientations. This flexibility allows for controlling the shape and arrangement of the fibers so that different structures (*i.e.* hollow, flat and ribbon shaped) can be fabricated depending on intended application purposes.

Nanofibers have many possible technological and commercial applications. They are used in tissue engineering, drug delivery, cancer diagnosis, lithium-air battery, optical sensors and air filtration.

Nanofibers were first produced via electrospinning more than four centuries ago. Beginning with the development of the electrospinning method, English physicist William Gilbert first documented the electrostatic attraction between liquids by preparing an experiment in which he observed a spherical water drop on a dry surface warp into a cone shape when it was held below an electrically charged amber. This deformation later came to be known as the Taylor cone. In 1882, English physicist Lord Rayleigh analyzed the unstable states of liquid droplets that were electrically charged, and noted that the liquid was ejected in tiny jets when equilibrium was established between the surface tension and electrostatic force. In 1887, British physicist Charles Vernon Boys published a manuscript about nanofiber development and production. In 1900, American inventor John Francis Cooley filed the first modern electrospinning patent.

Anton Formhals was the first person to attempt nanofiber production between 1934 and 1944 and publish the first patent describing the experimental production of nanofibers. In 1966, Harold Simons published a patent for a device that could produce thin and light nanofiber fabrics with diverse motifs.

Only at the end of the 20th century have the words electrospinning and nanofiber become common language among scientists and researchers. Electrospinning continues to be developed today.

Synthesis Methods

Many chemical and mechanical techniques for preparing nanofibers exist.

Electrospinning

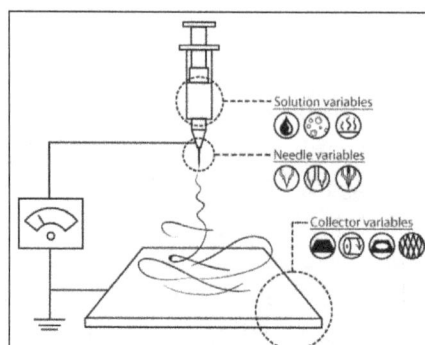

Diagram of a general set-up of electrospinning.

Electrospinning is the most commonly used method to fabricate nanofibers. The instruments necessary for electrospinning include a high voltage supplier, a capillary tube with a pipette or needle with a small diameter, and a metal collecting screen. One electrode is placed into the polymer solution and the other electrode is attached to the collector. An electric field is applied to the end of the capillary tube that contains the polymer solution held by its surface tension and forms a charge on the surface of the liquid. As the intensity of the electric field increases, the hemispherical surface of the fluid at the tip of the capillary tube elongates to form a conical shape known as the Taylor cone. A critical value is attained upon further increase in the electric field in which the repulsive electrostatic force overcomes the surface tension and the charged jet of fluid is ejected from the tip of the Taylor cone. The discharged polymer solution jet is unstable and elongates as a result, allowing the jet to become very long and thin. Charged polymer fibers solidifies with solvent evaporation. Randomly-oriented nanofibers are collected on the collector. Nanofibers can also be collected in a highly aligned fashion by using specialized collectors such as the rotating drum, metal frame, or a two-parallel plates system. Parameters such as jet stream movement and polymer concentration have to be controlled to produce nanofibers with uniform diameters and morphologies.

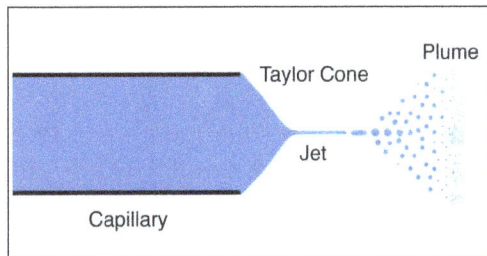

Taylor cone from which jet of polymer solution is ejected.

The electrospinning technique transforms many types of polymers into nanofibers. An electrospun nanofiber network resembles the extracellular matrix (ECM) well. This resemblance is a major advantage of electrospinning because it opens up the possibility of mimicking the ECM with regards to fiber diameters, high porosity, and mechanical properties. Electrospinning is being further developed for mass production of one-by-one continuous nanofibers.

Thermal-induced Phase Separation

Thermal-induced phase separation separates a homogenous polymer solution into a multi-phase system via thermodynamic changes. The procedure involves five steps: polymer dissolution, liquid-liquid or liquid-solid phase separation, polymer gelation, extraction of solvent from the gel with water, and freezing and freeze-drying under vacuum. Thermal-induced phase separation method is widely used to generate scaffolds for tissue regeneration.

The homogenous polymer solution in the first step is thermodynamically unstable and tends to separate into polymer-rich and polymer-lean phases under appropriate temperature. Eventually after solvent removal, the polymer-rich phase solidifies to form the matrix and the polymer-lean phase develops into pores. Next, two types of phase separation can be carried out on the polymer solution depending on the desired pattern. Liquid-liquid separation is usually used to form bicontinuous phase structures while solid-liquid phase separation is used to form crystal structures. The gelation step plays a crucial role in controlling the porous morphology of the nanofibrous matrices. Gelation is influenced by temperature, polymer concentration, and solvent properties.

Temperature regulates the structure of the fiber network: low gelation temperature results in formation of nanoscale fiber networks while high gelation temperature leads to the formation of a platelet-like structure. Polymer concentration affects fiber properties: an increase in polymer concentration decreases porosity and increases mechanical properties such as tensile strength. Solvent properties influence morphology of the scaffolds. After gelation, gel is placed in distilled water for solvent exchange. Afterwards, the gel is removed from the water and goes through freezing and freeze-drying. It is then stored in a desiccator until characterization.

Drawing

The drawing method makes long single strands of nanofibers one at a time. The pulling process is accompanied by solidification that converts the dissolved spinning material into a solid fiber. A cooling step is necessary in the case of melt spinning and evaporation of solvent in the case of dry spinning. A limitation, however, is that only a viscoelastic material that can undergo extensive deformations while possessing sufficient cohesion to survive the stresses developed during pulling can be made into nanofibers through this process.

Template Synthesis

The template synthesis method uses a nanoporous membrane template composed of cylindrical pores of uniform diameter to make fibrils (solid nanofiber) and tubules (hollow nanofiber). This method can be used to prepare fibrils and tubules of many types of materials, including metals, semiconductors and electronically conductive polymers. The uniform pores allow for control of the dimensions of the fibers so nanofibers with very small diameters can be produced through this method. However, a drawback of this method is that it cannot make continuous nanofibers one at a time.

Self-assembly

The self-assembly technique is used to generate peptide nanofibers and peptide amphiphiles. The method was inspired by the natural folding process of amino acid residues to form proteins with unique three-dimensional structures. The self-assembly process of peptide nanofibers involves various driving forces such as hydrophobic interactions, electrostatic forces, hydrogen bonding and van der Waals forces and is influenced by external conditions such as ionic strength and pH.

Polymer Materials

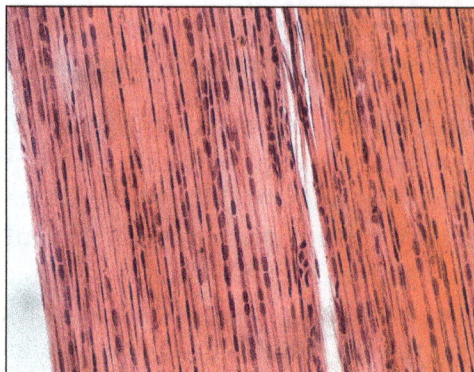

Collagen fibers in a cross-sectional area of dense connective tissue.

Due to their high porosity and large surface area-to-volume ratio, nanofibers are widely used to construct scaffolds for biological applications. Major examples of natural polymers used in scaffold production are collagen, cellulose, silk fibroin, keratin, gelatin and polysaccharides such as chitosan and alginate. Collagen is a natural extracellular component of many connective tissues. Its fibrillary structure, which varies in diameter from 50-500 nm, is important for cell recognition, attachment, proliferation and differentiation. Using type I collagen nanofibers produced via electrospinning, Shih et al. found that the engineered collagen scaffold showed an increase in cell adhesion and decrease in cell migration with increasing fiber diameter. Using silk scaffolds as a guide for growth for bone tissue regeneration, Kim et al. observed complete bone union after 8 weeks and complete healing of defects after 12 weeks whereas the control in which the bone did not have the scaffold displayed limited mending of defects in the same time period. Similarly, keratin, gelatin, chitosan and alginate demonstrate excellent biocompatibility and bioactivity in scaffolds.

However, cellular recognition of natural polymers can easily initiate an immune response. Consequently, synthetic polymers such as poly(lactic acid) (PLA), polycaprolactone (PCL), polyurethane (PU), poly(lactic-co-glycolic acid) (PLGA), poly(L-lactide) (PLLA), and poly(ethylene-co-vinylacetate) (PEVA) have been developed as alternatives for integration into scaffolds. Being biodegradable and biocompatible, these synthetic polymers can be used to form matrices with a fiber diameter within the nanometer range. Out of these synthetic polymers, PCL has generated considerable enthusiasm among researchers. PCL is a type of biodegradable polyester that can be prepared via ring-opening polymerization of ε-caprolactone using catalysts. It shows low toxicity, low cost and slow degradation. PCL can be combined with other materials such as gelatin, collagen, chitosan, and calcium phosphate to improve the differentiation and proliferation capacity. PLLA is another popular synthetic polymer. PLLA is well known for its superior mechanical properties, biodegradability and biocompatibility. It shows efficient cell migration ability due to its high spatial interconnectivity, high porosity and controlled alignment. A blend of PLLA and PLGA scaffold matrix has shown proper biomimetic structure, good mechanical strength and favorable bioactivity.

Applications

Tissue Engineering

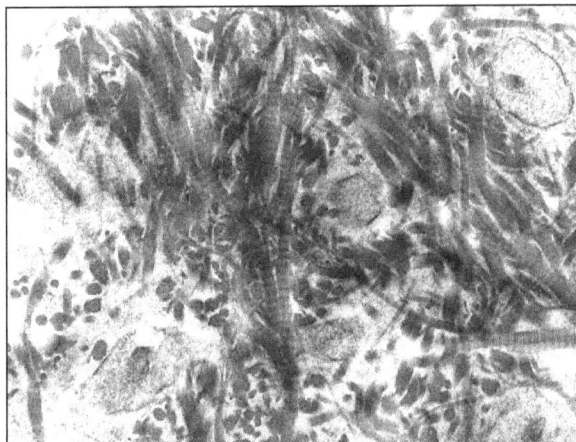

Bone matrix composed of collagen fibrils. Nanofiber
scaffolds are able to mimic such structure.

In tissue engineering, a highly porous artificial extracellular matrix is needed to support and guide cell growth and tissue regeneration. Natural and synthetic biodegradable polymers have been used to create such scaffolds.

Simon, in a 1988 NIH SBIR grant report, showed that electrospinning could be used to produced nano- and submicron-scale polystyrene and polycarbonate fibrous mats specifically intended for use as in vitro cell substrates. This early use of electrospun fibrous lattices for cell culture and tissue engineering showed that Human Foreskin Fibroblasts (HFF), transformed Human Carcinoma (HEp-2), and Mink Lung Epithelium (MLE) would adhere to and proliferate upon the fibers.

Nanofiber scaffolds are used in bone tissue engineering to mimic the natural extracellular matrix of the bones. The bone tissue is arranged either in a compact or trabecular pattern and composed of organized structures that vary in length from the centimeter range all the way to the nanometer scale. Nonmineralized organic component (i.e. type 1 collagen), mineralized inorganic component (i.e. hydroxyapatite), and many other noncollagenous matrix proteins (i.e. glycoproteins and proteoglycans) make up the nanocomposite structure of the bone ECM. The organic collagen fibers and the inorganic mineral salts provide flexibility and toughness, respectively, to ECM.

Although the bone is a dynamic tissue that can self-heal upon minor injuries, it cannot regenerate after experiencing large defects such as bone tumor resections and severe nonunion fractures because it lacks the appropriate template. Currently, the standard treatment is autografting which involves obtaining the donor bone from a non-significant and easily accessible site (i.e. iliac crest) in the patient own body and transplanting it into the defective site. Transplantation of autologous bone has the best clinical outcome because it integrates reliably with the host bone and can avoid complications with the immune system. But its use is limited by its short supply and donor site morbidity associated with the harvest procedure. Furthermore, autografted bones are avascular and hence are dependent on diffusion for nutrients, which affects their viability in the host. The grafts can also be resorbed before osteogenesis is complete due to high remodeling rates in the body. Another strategy for treating severe bone damage is allografting which transplants bones harvested from a human cadaver. However, allografts introduce the risk of disease and infection in the host.

Bone tissue engineering presents a versatile response to treat bone injuries and deformations. Nanofibers produced via electrospinning mimics the architecture and characteristics of natural extracellular matrix particularly well. These scaffolds can be used to deliver bioactive agents that promote tissue regeneration. These bioactive materials should ideally be osteoinductive, osteoconductive, and osseointegratable. Bone substitute materials intended to replace autologous or allogeneic bone consist of bioactive ceramics, bioactive glasses, and biological and synthetic polymers. The basis of bone tissue engineering is that the materials will be resorbed and replaced over time by the body's own newly regenerated biological tissue.

Tissue engineering is not only limited to the bone: a large amount of research is devoted to cartilage, ligament, skeletal muscle, skin, blood vessel, and neural tissue engineering as well.

Drug Delivery

Successful delivery of therapeutics to the intended target largely depends on the choice of the drug carrier. The criteria for an ideal drug carrier include maximum effect upon delivery of the drug to the target organ, evasion of the immune system of the body in the process of reaching the organ, retention of the therapeutic molecules from preparatory stages to the final delivery of the drug, and proper release of the drug for exertion of the intended therapeutic effect. Nanofibers are under study as a possible drug carrier candidate. Natural polymers such as gelatin and alginate make for good fabrication biomaterials for carrier nanofibers because of their biocompatibility and biodegradability that result in no harm to the tissue of the host and no toxic accumulation in the human body, respectively. Due to their cylindrical morphology, nanofibers possess a high surface area-to-volume ratio. As a result, the fibers possess high drug-loading capacity and may release therapeutic molecules over a large surface area. Whereas surface area to volume ratio can only be controlled by adjusting the radius for spherical vesicles, nanofibers have more degrees of freedom in controlling the ratio by varying both the length and the cross-sectional radius. This adjustability is important for their application in drug delivery system in which the functional parameters need to be precisely controlled.

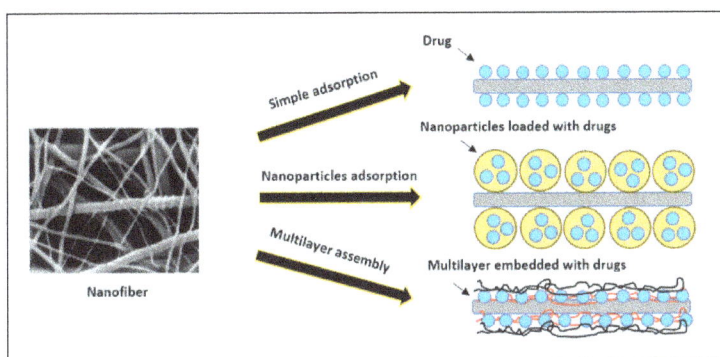

Drugs and biopolymers can be loaded onto nanofibers via simple adsorption, nanoparticles adsorption, and multilayer assembly.

Preliminary studies indicate that antibiotics and anticancer drugs may be encapsulated in electrospun nanofibers by adding the drug into the polymer solution prior to electrospinning. Surface-loaded nanofiber scaffolds are useful as adhesion barriers between internal organs and tissues post-surgery. Adhesion occurs during the healing process and can bring on complications such as chronic pain and reoperation failure.

Cancer Diagnosis

Although pathologic examination is the current standard method for molecular characterization in testing for the presence of biomarkers in tumors, these single-sample analyses fail to account for the diverse genomic nature of tumors. Considering the invasive nature, psychological stress, and the financial burden resulting from repeated tumor biopsies in patients, biomarkers that could be judged through minimally invasive procedures, such as blood draws, constitute an opportunity for progression in precision medicine.

Liquid biopsy is an option that is becoming increasingly popular as an alternative to solid tumor biopsy. This is simply a blood draw that contains circulating tumor cells (CTCs) which are shed into the bloodstream from solid tumors. Patients with metastatic cancer are more likely to have detectable CTCs in the bloodstream but CTCs also exist in patients with localized diseases. It has been found that the number of CTCs present in the bloodstream of patients with metastatic prostate and colorectal cancer is prognostic of the overall survival of tumors. CTCs also have been demonstrated to inform prognosis in earlier stages of the disease.

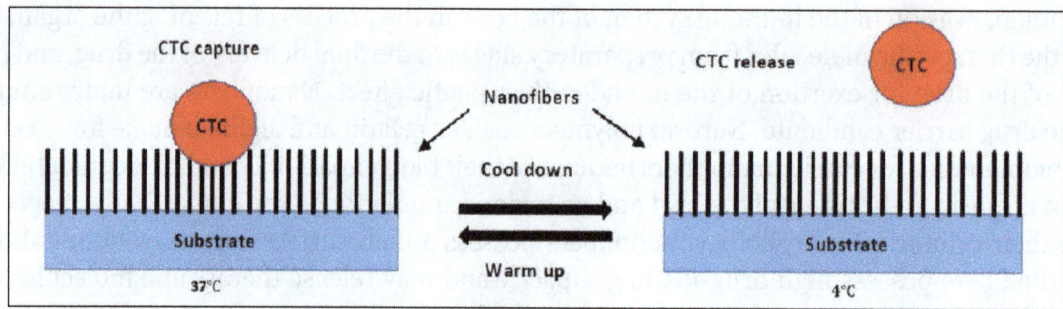

CTC capture and release mechanism of third generation Thermoresponsive Chip.

Recently, Ke et al. developed a NanoVelcro chip that captures the CTCs from the blood samples. When blood is passed through the chip, the nanofibers coated with protein antibodies bind to the proteins expressed on the surface of cancer cells and act like Velcro to trap CTCs for analysis. The NanoVelcro CTC assays underwent three generations of development. The first generation NanoVelcro Chip was created for CTC enumeration for cancer prognosis, staging, and dynamic monitoring. The second generation NanoVelcro-LCM was developed for single-cell CTC isolation. The individually isolated CTCs can be subjected to single-CTC genotyping. The third generation Thermoresponsive Chip allowed for CTC purification. The nanofiber polymer brushes undergo temperature-dependent conformational changes to capture and release CTCs.

Lithium-air Battery

Among many advanced electrochemical energy storage devices, rechargeable lithium-air batteries are of particular interest due to their considerable energy storing capacities and high power densities. As the battery is being used, lithium ions combine with oxygen from the air to form particles of lithium oxides, which attach to carbon fibers on the electrode. During recharging, the lithium oxides separate again into lithium and oxygen which is released back into the atmosphere. This conversion sequence is highly inefficient because there is significant voltage difference of more than 1.2 volts between the output voltage and the charging voltage of the battery meaning that approximately 30% of the electrical energy is lost as heat when the battery is charging. Also the large volume changes resulting from continuous conversion of oxygen between its gaseous and solid state puts stress on the electrode and limits its lifetime.

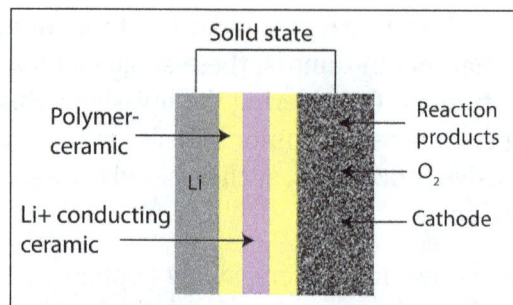

Schematic of a lithium-air battery. For the nanofiber-based lithium-air battery, the cathode would be made up of carbon nanofibers.

The performance of these batteries depends on the characteristics of the material that makes up the cathode. Carbon materials have been widely used as cathodes because of their excellent electrical conductivities, large surface areas, and chemical stability. Especially relevant for lithium-air batteries, carbon materials act as substrates for supporting metal oxides. Binder-free electrospun carbon nanofibers are particularly good potential candidates to be used in electrodes in lithium-oxygen batteries because they have no binders, have open macroporous structures, have carbons that support and catalyze the oxygen reduction reactions, and have versatility.

Zhu et al. developed a novel cathode that can store lithium and oxygen in the electrode they named nanolithia which is a matrix of carbon nanofibers periodically embedded with cobalt oxide. These cobalt oxides provide stability to the normally unstable superoxide-containing nanolithia. In this design, oxygen is stored as LiO_2 and does not convert between gaseous and solid forms during charging and discharging. When the battery is discharging, lithium ions in nanolithia and react with superoxide oxygen the matrix to form Li_2O_2, and Li_2O. The oxygen remains in its solid state as it transitions among these forms. The chemical reactions of these transitions provide electrical energy. During charging, the transitions occur in reverse.

Optical Sensors

Polymer optical fibers have generated increasing interest in recent years. Because of low cost, ease of handling, long wavelength transparency, great flexibility, and biocompatibility, polymer optical fibers show great potential for short-distance networking, optical sensing and power delivery.

Electrospun nanofibers are particularly well-suitable for optical sensors because sensor sensitivity increases with increasing surface area per unit mass. Optical sensing works by detecting ions and molecules of interest via fluorescence quenching mechanism. Wang et al. successfully developed nanofibrous thin film optical sensors for metal ion (Fe^{3+} and Hg^{2+}) and 2,4-dinitrotoluene (DNT) detection using the electrospinning technique.

Quantum dots show useful optical and electrical properties, including high optical gain and photochemical stability. A variety of quantum dots have been successfully incorporated into polymer nanofibers. Meng et al. showed that quantum dot-doped polymer nanofiber sensor for humidity detection shows fast response, high sensitivity, and long-term stability while requiring low power consumption.

Kelly et al. developed a sensor that warns first responders when the carbon filters in their respirators have become saturated with toxic fume particles. The respirators typically contain activated charcoal that traps airborne toxins. As the filters become saturated, chemicals begin to pass through and render the respirators useless. In order to easily determine when the filter is spent, Kelly and his team developed a mask equipped with a sensor composed of carbon nanofibers assembled into repeating structures called photonic crystals that reflect specific wavelengths of light. The sensors exhibit an iridescent color that changes when the fibers absorb toxins.

Air Filtration

Electrospun nanofibers are useful for removing volatile organic compounds (VOC) from the atmosphere. Scholten et al. showed that adsorption and desorption of VOC by electrospun nanofibrous membrane were faster than the rates of conventional activated carbon.

Airborne contamination in the personnel cabins of mining equipment is of concern to the mining workers, mining companies, and government agencies such as the Mine Safety and Health

Administration (MSHA). Recent work with mining equipment manufacturers and the MSHA has shown that nanofiber filter media can reduce cabin dust concentration to a greater extent compared to standard cellulose filter media.

Paints and protective coatings on furniture contain volatile organic compounds such as toluene and formaldehyde.

Nanofibers can be used in masks to protect people from viruses, bacteria, smog, dust, allergens and other particles. Filtration efficiency is at about 99.9% and the principle of filtration is mechanical. Particles in the air are bigger than pores in nanofiber web, but oxygen particles are small enough to pass through. Only one company put nanofibers to ordinary products in 2019. Products using nanofibers to filter air and protect peoples health can be as window screens, face masks, half masks or scarfs with nanofiber membrane. Campaign with nanofiber scarf shown on Kickstarter at the end of 2018.

Oil-water Separation

Nanofibers have the capabilities in oil–water separation, most particularly in sorption process when the material in use has the oleophilic and hydrophobic surfaces. These characteristic enable the nanofibers to be used as a tool to combat either oily waste-water from domestic household and industrial activities, or oily seawater due to the oil run down to the ocean from oil transportation activities and oil tank cleaning on a vessel.

Sportswear Textile

Sportswear textile with nanofiber membrane inside is based on the modern nanofiber technology where the core of the membrane consists of fibers with a diameter 1000× thinner than human hair. This extremely dense "sieve" with more than 2,5 billion of pores per square centimeter works much more efficiently with vapor removal and brings better level of water resistance. In the language of numbers, the nanofiber textile brings the following parameters:

- RET 1.0 vapor permeability and 10,000 mm water column (version preferring breathability).

- RET 4.8 vapor permeability and 30,000 mm water column (version preferring water resistance).

Nanofiber apparel and shoe membranes consist of polyurethane so its production is not harmful to nature. Membranes to sportswear made from nanofiber are recyclable.

GRAPHENE

Graphene is the name for a honeycomb sheet of carbon atoms. It is the building block for other graphitic materials (since a typical carbon atom has a diameter of about 0.33 nanometers, there are about 3 million layers of graphene in 1 mm of graphite).

Harder than diamond yet more elestic than rubber; tougher than steel yet lighter than aluminium. Graphene is the strongest known material.

To put this in perspective: if a sheet of cling film (like kitchen wrap film) had the same strength as a pristine monolayer of graphene, it would require the force exerted by a mass of 2000 kg, or a large car, to puncture it with a pencil.

Graphene possesses other amazing characteristics: Its high electron mobility is 100x faster than silicon; it conducts heat 2x better than diamond; its electrical conductivity is 13x better than copper; it absorbs only 2.3% of reflecting light; it is impervious so that even the smallest atom (helium) can't pass through a defect-free monolayer graphene sheet; and its high surface area of 2630 square meters per gram means that with less than 3 grams you could cover an entire soccer field (well, practically speaking you would need 6 grams, since 2630 m2/g is the surface area for both sides of a graphene sheet).

Graphene is the basic building block for other graphitic materials; it also represents a conceptually new class of materials that are only one atom thick, so-called two-dimensional (2D) materials (they are called 2D because they extends in only two dimensions: length and width; as the material is only one atom thick, the third dimension, height, is considered to be zero).

Graphene is also very attractive for the fabrication of mixed-dimensional van der Waals heterostructures that could be carried out through hybridizing graphene with 0D quantum dots or nanoparticles, 1D nanostructures such as nanowires or carbon nanotubes, or 3D bulk materials.

Graphene Properties

Electronic Properties

One of the reasons nanotechnology researchers working towards molecular electronics are so excited about graphene is its electronic properties – it is one of the best electrical conductors on Earth. The unique atomic arrangement of the carbon atoms in graphene allows its electrons to easily travel at extremely high velocity without the significant chance of scattering, saving precious energy typically lost in other conductors.

Scientists have found that graphene remains capable of conducting electricity even at the limit of nominally zero carrier concentration because the electrons don't seem to slow down or localize. The electrons moving around carbon atoms interact with the periodic potential of graphene's honeycomb lattice, which gives rise to new quasiparticles that have lost their mass, or rest mass(-so-called *massless Dirac fermions*). That means that graphene never stops conducting. It was also found that they travel far faster than electrons in other semiconductors.

Mechanical Properties

The impressive intrinsic mechanical properties of graphene, its stiffness, strength and toughness, are one of the reasons that make graphene stand out both as an individual material and as a reinforcing agent in composites. They are caused by the stability of the sp2 bonds that form the hexagonal lattice and oppose a variety of in-plane deformations.

Stiffness

The breaking force obtained experimentally and from simulation was almost identical and the experimental value of the second order elastic stiffness was equal to 340 ± 50 N m^{-1}. This value corresponds to a Young's modulus of 1.0 ± 0.1 TPa, assuming an effective thickness of 0.335 nm.

Strength

Defect-free, monolayer graphene is considered to be the strongest material ever tested with a strength of 42 N m^{-1}, which equates to an intrinsic strength of 130 GPa.

Toughness

Fracture toughness, which is a property very relevant to engineering applications, is one of the most important mechanical properties of graphene and was measured as a critical stress intensity factor of 4.0 ±0.6 MPa.

Research groups worldwide are working on the development of industrially manufacturable graphene sheets that have high strength and toughness in all sheet directions for diverse applications as graphene-based composites for vehicles, optoelectronics and neural implants.

A recent comsumer product example that exploits graphene's mechanical properties is the Momo Evo Graphene motorcycle helmet, developed by Italy's Momodesign and the Istituto Italiano di Tecnologia (IIT).

It is the first-ever graphene-infused carbon fiber helmet that capitalizes on the material's thin, strong and conductive, flexible and light characteristics to create a helmet that absorbs and dissipates impact better than your average helmet. It also disperses heat more efficiently, so it's cooler.

Another example is the Dassi Interceptor Graphene bike – the world's first graphene bicycle. Enhancing carbon fiber with graphene allows to make lighter, thinner tubes, that are stronger than regular carbon. That means an aero-shaped frame with none of the usual weight sacrifice. Thanks to its graphene reinforced frame, this bike is 30% lighter yet twice as strong and super stiff.

Graphene uses and Applications

Energy Storage and Solar Cells

Graphene-based nanomaterials have many promising applications in energy-related areas. Just some recent examples: Graphene improves both energy capacity and charge rate in rechargeable batteries; activated graphene makes superior supercapacitors for energy storage; graphene electrodes may lead to a promising approach for making solar cells that are inexpensive, lightweight and flexible; and multifunctional graphene mats are promising substrates for catalytic systems.

Researchers also have discovered a critical and unexpected relationship between the graphene's chemical/structural defectiveness as a host material for electrodes and its ability to suppress the growth of dendrites – branch-like filament deposits on the electrodes that can penetrate the barrier between the two halves of the battery and potentially cause electrical shorts, overheating and fires.

These examples highlight the four major energy-related areas where graphene will have an impact: solar cells, supercapacitors, graphene batteries, and catalysis for fuel cells.

Due to their excellent electron-transport properties and extremely high carrier mobility, graphene and other other direct bandgap monolayer materials such as transition-metal dichalcogenides (TMDCs) and black phosphorus show great potential to be used for low-cost, flexible, and highly efficient photovoltaic devices. They are the most promising materials for advanced solar cells.

An excellent review paper gives a brief overview of the recent research concerning chemical and thermal approaches toward the production of well-defined graphene-based nanomaterials and their applications in energy-related areas.

The authors note, however, that before graphene-based nanomaterials and devices find widespread commercial use, two important problems have to be solved: one is the preparation of graphene-based nanomaterials with well-defined structures, and the other is the controllable fabrication of these materials into functional devices.

Sensor Applications

Functionalized graphene holds exceptional promise for biological and chemical sensors. Already, researchers have shown that the distinctive 2D structure of graphene oxide (GO), combined with its superpermeability to water molecules, leads to sensing devices with an unprecedented speed.

Scientists have found that chemical vapors change the noise spectra of graphene transistors, allowing them to perform selective gas sensing for many vapors with a single device made of pristine graphene – no functionalization of the graphene surface required.

Quite a cool approach is to interface passive, wireless graphene nanosensors onto biomaterials via silk bioresorption as demonstrated by a graphene nanosensor tattoo on teeth monitors bacteria in your mouth.

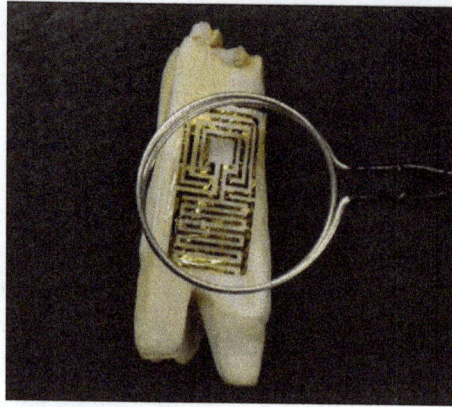

Optical image of the graphene wireless sensor
biotransferred onto the surface of a tooth.

Researchers also have begun to work with graphene foams – three-dimensional structures of interconnected graphene sheets with extremely high conductivity. These structures are very promising as gas sensors ("Graphene foam detects explosives, emissions better than today's gas sensors") and as biosensors to detect diseases.

Electronics Applications

Graphene has a unique combination of properties that is ideal for next-generation electronics, including mechanical flexibility, high electrical conductivity, and chemical stability. Combine this with inkjet printing and you get an inexpensive and scalable path for exploiting these properties in real-world technologies.

Transistors and Memory

Some of the most promising applications of graphene are in electronics (as transistors and interconnects), detectors (as sensor elements) and thermal management (as lateral heat spreaders). The first graphene field-effect transistors (FETs) – with both bottom and top gates – have already been demonstrated. At the same time, for any transistor to be useful for analog communication or digital applications, the level of the electronic low-frequency noise has to be decreased to an acceptable level.

Transistors on the basis of graphene are considered to be potential successors for the some silicon components currently in use. Due to the fact that an electron can move faster through graphene than through silicon, the material shows potential to enable terahertz computing.

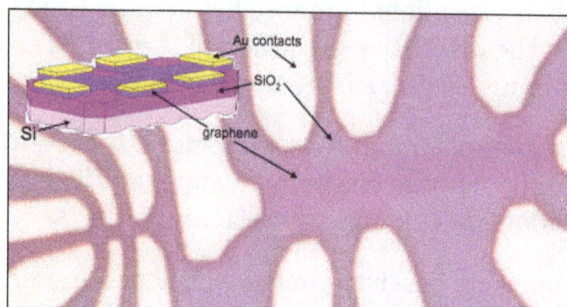

Graphene device: optical picture of a graphene device made out of a lithographically
cut graphene sheet on top of SiO_2, with gold electrodes and a doped Si back gate.

In the ultimate nanoscale transistor – dubbed a ballistic transistor – the electrons avoid collisions, i.e. there is a virtually unimpeded flow of current. Ballistic conduction would enable incredibly fast switching devices. Graphene has the potential to enable ballistic transistors at room temperature.

While graphene has the potential to revolutionize electronics and replace the currently used silicon materials, it does have an Achilles heel: pristine graphene is semi-metallic and lacks the necessary band gap to serve as a transistor. Therefore it is necessary to engineer band gaps in graphene.

Experiments have demonstrated the benefits of graphene as a platform for flash memory which show the potential to exceed the performance of current flash memory technology by utilizing the intrinsic properties of graphene.

Flexible, Stretchable and Foldable Electronics

Flexible electronics relies on bendable substrates and truly foldable electronics requires a foldable substrate with a very stable conductor that can withstand folding (i.e. an edge in the substrate at the point of the fold, which develops creases, and the deformation remains even after unfolding).

That means that, in addition to a foldable substrate like paper, the conductor that is deposited on this substrate also needs to be foldable. To that end, researchers have demonstrated a fabrication process for foldable graphene circuits based on paper substrates.

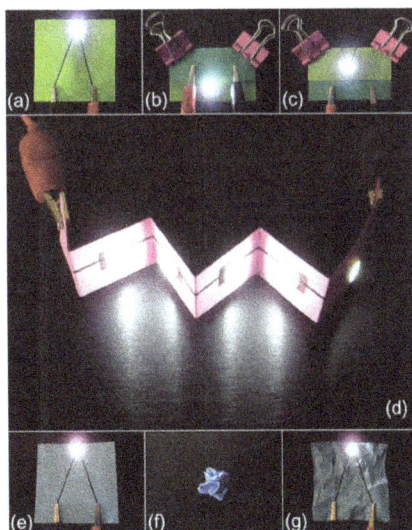

The figure shows, photographs of applications. a,b,c) Operation of a LED chip with graphene circuits on a paper substrate under -180° folding and 180° folding. d) Array of LED chips on a three-dimensional circuit board including negative and positive angle folding. e,f,g) Operation of a LED chip on the paper-based circuit board before and after crumpling.

Graphene's remarkable conductivity, strength and elasticity has also made it a promising choice for stretchable electronics – a technology that aims to produce circuits on flexible plastic substrates for applications like bendable solar cells or robotic-like artificial skin.

Scientists have devised a chemical vapor deposition (CVD) method for turning graphene sheets into porous three-dimensional foams with extremely high conductivity. By permeating this foam with a siloxane-based polymer, the researchers have produced a composite that can be twisted, stretched and bent without harming its electrical or mechanical properties.

Photodetectors

Researchers have demonstrated that graphene can be used for telecommunications applications and that its weak and universal optical response might be turned into advantages for ultrafast photonics applications. They also found that graphene could be potentially exploited as a saturable absorber with wide optical response ranging from ultra-violet, visible, infrared to terahertz.

There is a very strong research interest in using graphene for applications in optoelectronics. Graphene-based photodetectors have been realized before and graphene's suitability for high bandwidth photodetection has been demonstrated in a 10 GBit/s optical data link.

One novel approach is based on the integration of graphene into an optical microcavity. The increased electric field amplitude inside the cavity causes more energy to be absorbed, leading to a significant increase of the photoresponse.

Coatings

Coating objects with graphene can serve different purposes. For instance, researchers have now shown that it is possible to use graphene sheets to create a superhydrophobic coating material that shows stable superhydrophobicity under both static as well as dynamic (droplet impact) conditions, thereby forming extremely water repelling structures.

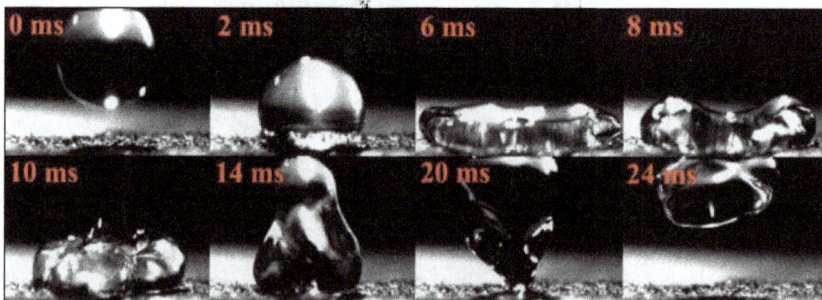

Snapshots of a water droplet impacting the surface of the Teflon coated graphene foam. The impact velocity just prior to the droplet striking the surface was ~76 cm/sec. The sequence of snapshots shows the deformation time history of the droplet upon impact. The droplet spreads, then retracts and successfully rebounds off the surface. The coefficient of restitution (i.e. ratio of droplet impacting velocity to ejecting velocity) is ~0.37 for the Teflon coated foam.

Graphene also is the world's thinnest known coating for protecting metals against corrosion. It was found that graphene, whether made directly on copper or nickel or transferred onto another metal, provides protection against corrosion.

Researchers demonstrated the use of graphene as a transparent conductive coating for photonic devices and show that its high transparency and low resistivity make this two-dimensional crystal ideally suitable for electrodes in liquid crystal devices (LCDs).

Another novel coating application useful for researchers is the the fabrication of polymeric AFM probes covered by monolayer graphene to improving AFM probe performance.

GRAPHANE

Graphane is a two-dimensional polymer of carbon and hydrogen with the formula unit $(CH)_n$ where n is large. Graphane should not be confused with graphene, a two-dimensional form of carbon alone. Graphane is a form of hydrogenated graphene. Graphane's carbon bonds are in sp^3 configuration, as opposed to graphene's sp^2 bond configuration, thus graphane is a two-dimensional analog of cubic diamond.

Structure

The structure was found, using a cluster expansion method, as the most stable of all the possible hydrogenation ratios of graphene in 2003. In 2007, researchers found that the compound is more stable than other compounds containing carbon and hydrogen, such as benzene, cyclohexane and polyethylene. This group named the predicted compound graphane, because it is the fully saturated version of graphene. The compound is an insulator. Chemical functionalization of graphene with hydrogen may be a suitable method to open a band gap in graphene.

P-doped graphane is proposed to be a high-temperature BCS theory superconductor with a T_c above 90 K.

Any disorder in hydrogenation conformation tends to contract the lattice constant by about 2.0%.

Variants

Partial hydrogenation leads to hydrogenated graphene rather than (fully hydrogenated) graphane. Such compounds are usually named as "graphane-like" structures. Graphane and graphane-like structures can be formed by electrolytic hydrogenation of graphene or few-layer graphene or high-oriented pyrolytic graphite. In the last case mechanical exfoliation of hydrogenated top layers can be used.

Hydrogenation of graphene on substrate affects only one side, preserving hexagonal symmetry. One-sided hydrogenation of graphene is possible due to the existence of ripplings. Because the latter are distributed randomly, the obtained material is disordered in contrast to two-sided graphane. Annealing allows the hydrogen to disperse, reverting to graphene. Simulations revealed the underlying kinetic mechanism.

Density functional theory calculations suggested that hydrogenated and fluorinated forms of other group IV (Si, Ge and Sn) nanosheets present properties similar to graphane.

Potential Applications

Graphane has been proposed for hydrogen storage. Hydrogenation decreases the dependence of the lattice constant on temperature, which indicates a possible application in precision instruments.

SILICENE

Silicene is the thinnest possible form of silicon, and could have promising applications in micro-electronics if it can be manufactured commercially. Silicene is the silicon equivalent of graphene, the wonder material that has gripped the world of nanoscience since 2004.

Properties of Silicene

Two-dimensional silicene is not completely planar - like graphene it consists of a hexagonal pattern of atoms, but in silicene the hexagonal rings include puckering distortions due to a "chair" configuration. Hence the surface seems to have ordered ripples.

Silicene can be converted to silicane by addition of hydrogen in an exothermic chemical reaction. This process could lead to applications in hydrogen storage.

As silicene is composed of silicon, it is compatible with present semiconductor manufacturing techniques. It's edges do not show oxygen reactivity, which suggests it will be stable enough to use in electronics. Some experiments show that silicene has an electronic structure which is analogous to that of graphene, which allow electrons to move through the material as though they had zero mass. Not all the evidence agrees with this, however, and the exact electronic properties have yet to be determined.

Comparison to Graphene

Graphene is often referred to as a wonder material. It has superb electron mobility, and is so strong that it is possible to hold a single atomic sheet. It's density is around a million times that of copper at room temperature.

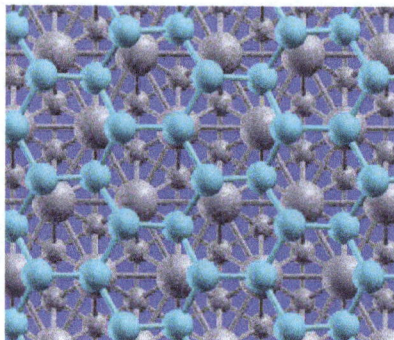

Diagram of the structure of silicene on a ZrB2 substrate.

If silicene can be produced reliably, it could be even more exciting than graphene, since it will be more directly compatible with present silicon-based electronics, allowing its benefits to be more easily exploited. According to the European research team, silicene behaved in a similar way to graphene in terms of it's electronic properties - they can move like particles without mass, traveling through the lattice at the speed of light on a continuous range of energy levels.

Graphene does not have any bandgap - it is a semi-metal, which means it has very good conductivity. However, a non-zero bandgap is what makes a material a semiconductor, and what allows semiconductors to be used in electronic components like diodes and transistors. Silicene does have a bandgap, hence making it suitable to be used in novel transistors and other.

Recent work used simple vapor deposition techniques to grow a silicon layer one atom thick on a silver crystal surface. Observing the chemical structure, the dimensions and electronics of the produced material, the researchers concluded that this must be silicene. The structural parameters such as lengths and bond angles matched density functional theory (DFT) calculations perfectly.

The interlayer silicene coupling is very strong when compared to graphite, according the European research. However, results reported by the Japanese team do not show an identical behavior. A physicist from the Italian National Research Council in Rome, Paola De Padova, who was involved in the original European study says that the absence of the Dirac cone leaves room for doubt that silicene is very similar to graphene. A senior author of the Japanese research team felt that it is better not to define silicene too rigidly since the work is still it its early stages.

According to an associate professor of inorganic chemistry at the University of Nottingham, UK, silicene may be more convenient to be integrated into electronic devices when compared to graphene.

An Aix-Marseille group tried creating silicene in 2010 but they could not find considerable evidence as to its presence. According to Michel Houssa of the Catholic University of Leuven (KUL) in Belgium, this study proved that it is possible to grow silicene on silver. Angle-resolved photo-emission spectroscopy was used and it was observed that silver sheets synthesized on silver surface show electrical characteristics similar to graphene. A Fermi-velocity of vF = 1.3×106 ms^{-1} was determined for these silicene sheets that is comparable to graphene. Silicene has also been reported to grow on a ZrB$_2$ substrate. He added that it will truly be a challenge to grow silicene on insulating substrates so that more is understood about its electrical properties and study how they can be used to build electronic devices.

Applications of Silicene

Density functional theory calculations that were done in 2011 showed that silicene clusters will be suitable for FET applications.

Since silicene has a band gap unlike graphene, it will be possible to develop nanoscale semiconductors with a wide range of applications including light emitting diodes and microprocessor chips.

NANOSHEET

A nanosheet is a two-dimensional nanostructure with thickness in a scale ranging from 1 to 100 nm.

A typical example of a nanosheet is graphene, the thinnest two-dimensional material (0.34 nm) in the world. It consists of a single layer of carbon atoms with hexagonal lattices.

Examples and Applications

As of 2017 Silicon nanosheets are being used to prototype future generations of small (5 nm) transistors.

Carbon nanosheets (from hemp) may be an alternative to graphene as electrodes in supercapacitors.

Synthesis

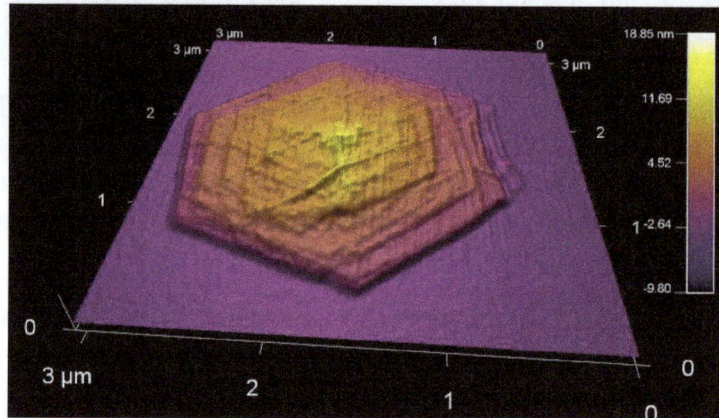

3D AFM topography image of multilayered palladium nanosheet on silicon wafer.

The most commonly used nanosheet synthesis methods use a bottom-up approach, e.g., pre-organization and polymerization at interfaces like Langmuir–Blodgett films, solution phase synthesis and chemical vapor deposition (CVD). For example, CdTe (cadmium telluride) nanosheets could be synthesized by precipitating and aging CdTe nanoparticles in deionized water. The formation of free-floating CdTe nanosheets was due to directional hydrophobic attraction and anisotropic electrostatic interactions caused by dipole moment and small positive charges. Molecular simulations through a coarse-grained model with parameters from semi-empirical quantum mechanics calculations can be used to prove the experimental process.

Ultrathin single-crystal PbS (lead sulfur) sheets with micro scale in x-, y- dimensions can be obtained using a hot colloidal synthesis method. Compounds with linear chloroalkanes like 1,2-dichloroethane containing chlorine were used during the formation of PbS sheets. PbS ultrathin sheets probably resulted from the oriented attachment of the PbS nanoparticles in a two-dimensional fashion. The highly reactive facets were preferentially consumed in the growth process that led to the sheet-like PbS crystal growth.

Nanosheets can also be prepared at room temperature. For instance, hexagonal PbO (lead oxide)) nanosheets were synthesized using gold nanoparticles as seeds under room temperature. The size of the PbO nanosheet can be tuned by gold NPs and Pb^{2+}concentration in the growth solution. No organic surfactants were employed in the synthesis process. Oriented attachment, in which the sheets form by aggregation of small nanoparticles that each has a net dipole moment, and ostwald ripening are the two main reasons for the formation of the PbO nanosheets. The same process was observed for iron sulfide nanoparticles.

Carbon nanosheets have been produced using industrial hemp bast fibres with a technique that involves heating the fibres at over 35 °F (18 °C) for 24 hours. The result is then subjected to intense heat causing the fibers to exfoliate into a carbon nanosheet. This has been used to create an electrode for a supercapacitor with electrochemical qualities 'on a par with' devices made using graphene.

Metal nanosheets have also been synthesized from solution-based method by reducing metal precursors, including palladium, rhodium, and gold.

Boron Nitride Nanosheet

Boron nitride nanosheet is a two-dimensional crystalline form of the hexagonal boron nitride (h-BN), which has a thickness of one to few atomic layers. It is similar in geometry to its all-carbon analog graphene, but has very different chemical and electronic properties – contrary to the black and highly conducting graphene, BN nanosheets are electrical insulators with a band gap of ~5.9 eV, and therefore appear white in color.

Uniform monoatomic BN nanosheets can be deposited by catalytic decomposition of borazine at a temperature ~1100 °C in a chemical vapor deposition setup, over substrate areas up to about 10 cm². Owing to their hexagonal atomic structure, small lattice mismatch with graphene (~2%), and high uniformity they are used as substrates for graphene-based devices.

Structure

BN nanosheets consist of sp^2-conjugated boron and nitrogen atoms that form a honeycomb structure. They contain two different edges: armchair and zig-zag. The armchair edge consists of either boron or nitrogen atoms, while the zig-zag edge consists of alternating boron and nitrogen atoms. These 2D structures can stack on top of each other and are held by Van der Waals forces to form few-layer boron nitride nanosheets. In these structures, the boron atoms of one sheet are positioned on top or below the nitrogen atoms due to electron-deficient nature of boron and electron-rich nature of nitrogen.

Synthesis

CVD

Chemical vapor deposition is the most common method to produce BN nanosheets because it is a well-established and highly controllable process that yields high-quality material over areas exceeding 10 cm². There is a wide range of boron and nitride precursors for CVD synthesis, such as borazine, and their selection depends on toxicity, stability, reactivity, and the nature of the CVD method.

Mechanical Cleavage

Mechanical cleaving methods of boron nitride use shear forces to break the weak van der Waals interactions between the BN layers. Cleaved nanosheets have low defect densities and retain the lateral size of the original substrate. Inspired by its use in the isolation of graphene, micromechanical cleavage, also known as the Scotch-tape method, has been used to consistently isolate few-layer and monolayer boron nitride nanosheets by subsequent exfoliation of the starting material with adhesive tape. The disadvantage of this technique is that it is not scalable for large-scale production.

A typical electron micrograph of BN nanosheets prepared by ball milling (scale bar 50 nm).

Boron nitride sheets can be also exfoliated by ball milling, where shear forces are applied on the face of bulk boron nitride by rolling balls. This technique yields large quantities of low-quality material with poor control over its properties.

Unzipping of Boron Nitride Nanotubes

BN nanosheets can be synthesized by the unzipping boron nitride nanotubes via potassium intercalation or etching by plasma or an inert gas. Here the intercalation method has a relatively low yield as boron nitride is resistive to the effects of intercalants. In situ unzipping of boron nitride nanotubes to nanoribbons was achieved by Li et al.

Solvent Exfoliation and Sonication

Solvent exfoliation is often used in tandem with sonication to isolate large quantities of boron nitride nanosheets. Polar solvents such as isopropyl alcohol and DMF are more effective in exfoliating boron nitride layers than nonpolar solvents because these solvents possess a similar surface energy to the surface energy of boron nitride nanosheets. Combinations of different solvents also exfoliate boron nitride better than individual solvents. Many solvents suitable for BN exfoliation are rather toxic and expensive, but they can be replaced by water and isopropyl alcohol without significantly sacrificing the yield.

Chemical Functionalization and Sonication

Chemical functionalization of boron nitride involves attaching molecules onto the outer and inner layers of bulk boron nitride. There are three types of BN functionalization: covalent, ionic and or non-covalent. Layers are exfoliated by placing the functionalized BN into a solvent and allowing the solvation force between the attached groups and the solvent to break the van der Waal forces between BN layers. This method is slightly different from solvent exfoliation, which relies on the similarities between the surface energies of the solvent and boron nitride layers.

Solid State Reactions

Heating a mixture of boron and nitrogen precursors, such as boric acid and urea, can produce boron nitride nanosheets. The number of layers in these nanosheets was controlled by temperature (ca. 900 °C) and the urea content.

Properties and Applications

- Mechanical properties: Monolayer boron nitride has an average Young's modulus of 0.865 TPa and fracture strength of 70.5 GPa. In contrast to graphene, whose strength decreases dramatically with increased thickness, few-layer boron nitride sheets have a strength similar to that of monolayer boron nitride.

- Thermal conductivity: The thermal conductivity of atomically thin boron nitride is one of the highest among semiconductors and electrical insulators; it increases with reduced thickness due to less intra-layer coupling.

- Thermal stability: The air stability of graphene shows a clear thickness dependence: monolayer graphene is reactive to oxygen at 250 °C, strongly doped at 300 °C, and etched at 450 °C; in contrast, bulk graphite is not oxidized until 800 °C. Atomically thin boron nitride has much better oxidation resistance than graphene. Monolayer boron nitride is not oxidized till 700 °C and can sustain up to 850 °C in air; bilayer and trilayer boron nitride nanosheets have slightly higher oxidation starting temperatures. The excellent thermal stability, high impermeability to gas and liquid, and electrical insulation make atomically thin boron nitride potential coating materials for preventing surface oxidation and corrosion of metals and other two-dimensional (2D) materials, such as black phosphorus.

- Better surface adsorption: Atomically thin boron nitride has been found to have better surface adsorption capabilities than bulk hexagonal boron nitride. According to theoretical and experimental studies, atomically thin boron nitride as an adsorbent experiences conformational changes upon surface adsorption of molecules, increasing adsorption energy and efficiency. The synergic effect of the atomic thickness, high flexibility, stronger surface adsorption capability, electrical insulation, impermeability, high thermal and chemical stability of BN nanosheets can increase the Raman sensitivity by up to two orders, and in the meantime attain long-term stability and extraordinary reusability not achievable by other materials.

- Dielectric properties: Atomically thin hexagonal boron nitride is an excellent dielectric substrate for graphene, molybdenum disulphide (MoS_2), and many other 2D material-based electronic and photonic devices. As shown by electric force microscopy (EFM) studies, the electric field screening in atomically thin boron nitride shows a weak dependence on thickness, which is in line with the smooth decay of electric field inside few-layer boron nitride revealed by the first-principles calculations.

- Raman characteristics: Raman spectroscopy has been a useful tool to study a variety of 2D materials, and the Raman signature of high-quality atomically thin boron nitride was first reported by Gorbachev et al. and Li et al. However, the two reported Raman results of monolayer boron nitride did not agree with each other. Cai et al. conducted systematic experimental and theoretical studies of the intrinsic Raman spectrum of atomically thin boron nitride. They reveal that, in absence of interaction with a substrate, atomically thin boron nitride has a G-band frequency similar to that of bulk hexagonal boron nitride, but strain induced by the substrate can cause Raman shifts. Nevertheless, the Raman intensity of G-band can be used to estimate layer thickness and sample quality.

BN nanosheets are electrical insulators and have a wide band gap of ~5.9 eV, which can be changed by the presence of Stone–Wales defects within the structure, by doping or functionalization, or by changing the number of layers. Owing to their hexagonal atomic structure, small lattice mismatch with graphene (~2%), and high uniformity, BN nanosheets are used as substrates for graphene-based devices. BN nanosheets are also excellent proton conductors. Their high proton transport rate, combined with the high electrical resistance, may lead to applications in fuel cells and water electrolysis.

NANOMESH

The nanomesh is a inorganic nanostructured two-dimensional material, similar to graphene. It was discovered in 2003 at the University of Zurich, Switzerland .

It consists of a single layer of boron (B) and nitrogen (N) atoms, which forms by self-assembly into a highly regular mesh after high-temperature exposure of a clean rhodium or ruthenium surface to borazine under ultra-high vacuum.

The nanomesh looks like an assembly of hexagonal pores at the nanometer (nm) scale. The distance between two pore centers is only 3.2 nm, whereas each pore has a diameter of about 2 nm and is 0.05 nm deep. The lowest regions bind strongly to the underlying metal, while the wires (highest regions) are only bound to the surface through strong cohesive forces within the layer itself.

The boron nitride nanomesh is not only stable under vacuum, air and some liquids, but also up to temperatures of 796 °C (1070 K). In addition it shows the extraordinary ability to trap molecules and metallic clusters, which have similar sizes to the nanomesh pores, forming a well-ordered array. These characteristics promise interesting applications of the nanomesh in areas like nanocatalysis, surface functionalisation, spintronics, quantum computing and data storage media like hard drives.

Structure

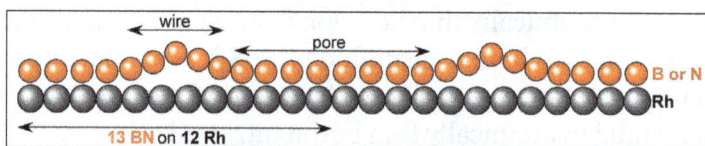

Cross-section of nanomesh on rhodium showing pore and wire regions.

h-BN nanomesh is a single sheet of hexagonal boron nitride, which forms on substrates like rhodium Rh(111) or ruthenium Ru(0001) crystals by a self-assembly process. The unit cell of the h-BN nanomesh consists of 13x13 BN or 12x12 Rh atoms with a lattice constant of 3.2 nm. In a cross-section it means that 13 boron or nitrogen atoms are sitting on 12 rhodium atoms. This implies a modification of the relative positions of each BN towards the substrate atoms within a unit cell, where some bonds are more attractive or repulsive than other (site selective bonding), what induces the corrugation of the nanomesh.

The nanomesh corrugation amplitude of 0.05 nm causes a strong effect on the electronic structure, where two distinct BN regions are observed. They are easily recognized in the lower right figure, which is a scanning tunneling microscopy (STM) measurement, as well as in the lower left figure representing a theoretical calculation of the same area. A strongly bounded region assigned to the pores is visible in blue in the left figure below (center of bright rings in the figure below) and a weakly bound region assigned to the wires appears yellow-red in the figure below (area in-between rings in the figure below).

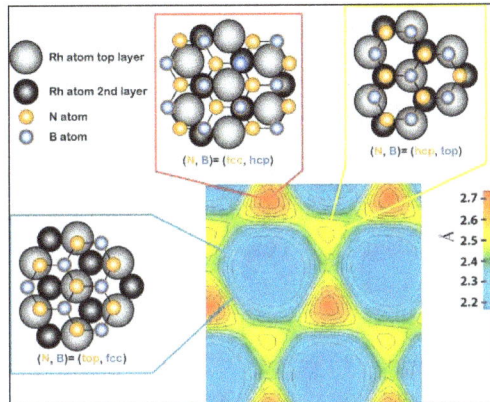

The theoretical calculation of the same area, where the N height relative to the underlying substrate is given. The exact arrangement of Rh, N and B atoms is given for three different areas (blue: pores, yellow-red: wires).

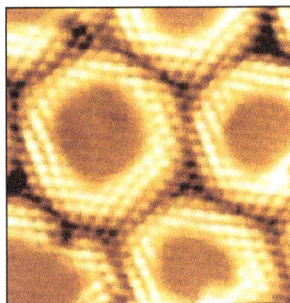

The boron nitride nanomesh measured by STM at 77 K, where each "ball" represents one N atom. The center of each ring corresponds to the center of the pores.

Properties

The nanomesh is stable under a wide range of environments like air, water and electrolytes among others. It is also temperature resistant since it does not decompose in temperatures up to 1275K under a vacuum. In addition to these exceptional stabilities, the nanomesh shows the extraordinary ability to act as a scaffold for metallic nanoclusters and to trap molecules forming a well-ordered array. In the case of gold (Au), its evaporation on the nanomesh leads to formation of well-defined round Au nanoparticles, which are centered at the nanomesh pores.

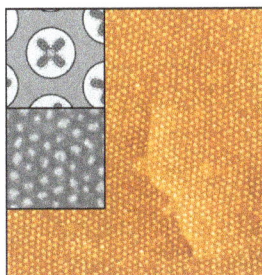

Naphthalocyanine molecules evaporated onto the nanomesh. They only adsorb in pores, forming a well-defined pattern.

The STM figure on the right shows Naphthalocyanine (Nc) molecules, which were vapor-deposited onto the nanomesh. These planar molecules have a diameter of about 2 nm, whose size is comparable to that of the nanomesh pores. It is spectacularly visible how the molecules form a well-ordered array with the periodicity of the nanomesh (3.22 nm). The lower inset shows a region of this substrate with higher resolution, where individual molecules are trapped inside the pores. In addition, the molecules seem to keep their native conformation, what means that their functionality is kept, which is nowadays a challenge in nanoscience.

Such systems with wide spacing between individual molecules/clusters and negligible intermolecular interactions might be interesting for applications such as molecular electronics and memory elements, in photochemistry or in optical devices.

Preparation and Analysis

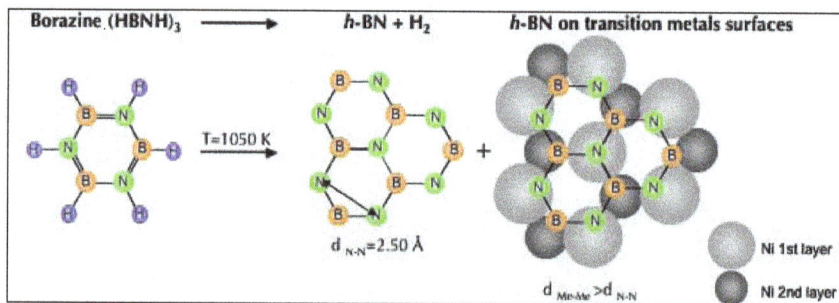

Decomposition of borazine on transition metal surfaces.

Well-ordered nanomeshes are grown by thermal decomposition of borazine $(HBNH)_3$, a colorless substance that is liquid at room temperature. The nanomesh results after exposing the atomically clean Rh(111) or Ru(0001) surface to borazine by chemical vapor deposition (CVD).

The substrate is kept at a temperature of 796 °C (1070 K) when borazine is introduced in the vacuum chamber at a dose of about 40 L (1 Langmuir = 10^{-6} torr sec). A typical borazine vapor pressure inside the ultrahigh vacuum chamber during the exposure is 3×10^{-7} mbar.

After cooling down to room temperature, the regular mesh structure is observed using different experimental techniques. Scanning tunneling microscopy (STM) gives a direct look on the local real space structure of the nanomesh, while low energy electron diffraction (LEED) gives information about the surface structures ordered over the whole sample. Ultraviolet photoelectron spectroscopy (UPS) gives information about the electronic states in the outermost atomic layers of a sample, i.e. electronic information of the top substrate layers and the nanomesh.

CORE–SHELL SEMICONDUCTOR NANOCRYSTAL

Core–shell semiconducting nanocrystals (CSSNCs) are a class of materials which have properties intermediate between those of small, individual molecules and those of bulk, crystalline semiconductors. They are unique because of their easily modular properties, which are a result of their size. These nanocrystals are composed of a quantum dot semiconducting core material and a shell of a distinct semiconducting material. The core and the shell are typically composed of type

II–VI, IV–VI, and III–V semiconductors, with configurations such as CdS/ZnS, CdSe/ZnS, CdSe/CdS, and InAs/CdSe (typical notation is: core/shell) Organically passivated quantum dots have low fluorescence quantum yield due to surface related trap states. CSSNCs address this problem because the shell increases quantum yield by passivating the surface trap states. In addition, the shell provides protection against environmental changes, photo-oxidative degradation, and provides another route for modularity. Precise control of the size, shape, and composition of both the core and the shell enable the emission wavelength to be tuned over a wider range of wavelengths than with either individual semiconductor. These materials have found applications in biological systems and optics.

Colloidal semiconductor nanocrystals, which are also called quantum dots (QDs), consist of ~1–10 nm diameter semiconductor nanoparticles that have organic ligands bound to their surface. These nanomaterials have found applications in nanoscale photonic, photovoltaic, and light-emitting diode (LED) devices due to their size-dependent optical and electronic properties. Quantum dots are popular alternatives to organic dyes as fluorescent labels for biological imaging and sensing due to their small size, tuneable emission, and photostability.

The luminescent properties of quantum dots arise from exciton decay (recombination of electron hole pairs) which can proceed through a radiative or nonradiative pathway. The radiative pathway involves electrons relaxing from the conduction band to the valence band by emitting photons with wavelengths corresponding to the semiconductor's bandgap. Nonradiative recombination can occur through energy release via phonon emission or Auger recombination. In this size regime, quantum confinement effects lead to a size dependent increasing bandgap with observable, quantized energy levels. The quantized energy levels observed in quantum dots lead to electronic structures that are intermediate between single molecules which have a single HOMO-LUMO gap and bulk semiconductors which have continuous energy levels within bands.

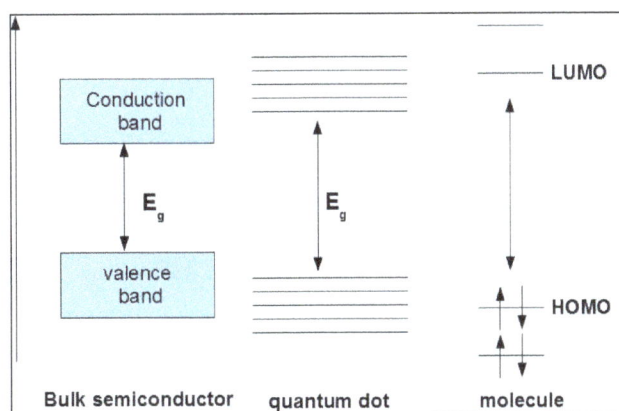

Semiconductor nanocrystals generally adopt the same crystal structure as their extended solids. At the surface of the crystal, the periodicity abruptly stops, resulting in surface atoms having a lower coordination number than the interior atoms. This incomplete bonding (relative to the interior crystal structure) results in atomic orbitals that point away from the surface called "dangling orbitals" or unpassivated orbitals. Surface dangling orbitals are localized and carry a slight negative or positive charge. Weak interaction among the inhomogeneous charged energy states on the surface has been hypothesized to form a band structure. If the energy of the dangling orbital band is within the semiconductor bandgap, electrons and holes can be trapped at the crystal surface. For example, in CdSe quantum dots, Cd dangling orbitals act as electron traps while Se dangling orbitals act as hole traps. Also, surface defects in the crystal structure can act as charge carrier traps.

Charge carrier trapping on QDs increases the probability of non-radiative recombination, which reduces the fluorescence quantum yield. Surface-bound organic ligands are typically used to coordinate to surface atoms having reduced coordination number in order to passivate the surface traps. For example, tri-n-octylphosphine oxide (TOPO) and trioctylphospine(TOP) have been used to control the growth conditions and passivate the surface traps of high quality CdSe quantum dots. Although this method provides narrow size distributions and good crystallinity, the quantum yields are ~5–15%. Alkylamines have been incorporated into the TOP/TOPO synthetic method to increase the quantum yields to ~50%.

The main challenge in using organic ligands for quantum dot surface trap passivation is the difficulty in simultaneously passivating both anionic and cationic surface traps. Steric hindrance between bulky organic ligands results in incomplete surface coverage and unpassivated dangling orbitals. Growing epitaxial inorganic semiconductor shells over quantum dots inhibits photo-oxidation and enables passivation of both anionic and cationic surface trap states. As photogenerated charge carriers are less likely to be trapped, the probability for excitons to decay through the radiative pathway increases. CdSe/CdS and ZnSe/CdSe nanocrystals have been synthesized that exhibit 85% and 80–90% quantum yield, respectively.

Core–shell semiconductor nanocrystal architecture was initially investigated in the 1980s, followed by a surge of publications on synthetic methods the 1990s.

Classification of Core–shell Semiconductor Nanocrystals

Core–shell semiconductor nanocrystal properties are based on the relative conduction and valence band edge alignment of the core and the shell. In type I semiconductor heterostructures, the electron and holes tend to localize within the core. In type II heterostructures, one carrier is localized in the shell while the other is localized in the core.

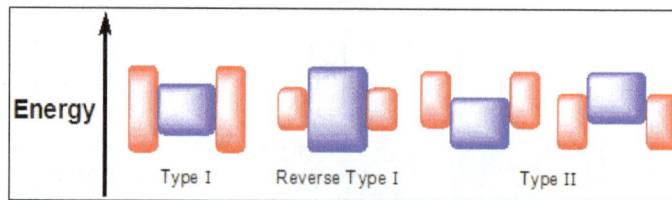

The three types of core–shell nanocrystals. The upper and lower edges represent the upper and lower energy edges of the core (blue) and the shell (red).

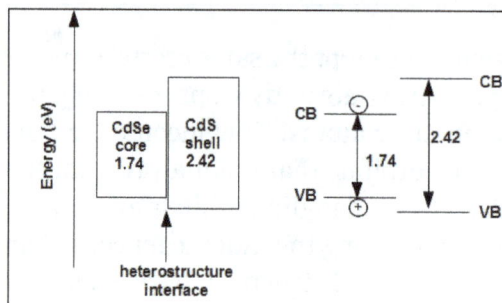

Type I

In a Type I CSSNC, the bandgap of the core is smaller than that of the shell. Both the conduction and valence band edges of the core lie within the bandgap of the shell, which confines both electrons and holes in the core. This can be seen in figure, where the electron and hole of an exciton at the CdSe (bandgap:1.74 eV)/CdS (bandgap:2.42 eV) interface occupy energy states within the CdSe core, which corresponds to the lowest available energy separation. The emission wavelength due to radiative electron-hole recombination within the core is slightly redshifted compared to uncoated CdSe.

Examples: CdSe/CdS, CdSe/ZnS, InAs/CdSe and ZnO/MgO

Reverse Type I

In the reverse type I configuration, the core has a wider bandgap than the shell, and the conduction and valence band edges of the shell lie within those of the core. The lowest available exciton energy separation occurs when the charge carriers are localized in the shell. Changing the shell thickness tunes the emission wavelength.

Examples: CdS/HgS, CdS/CdSe, ZnSe/CdSe and MgO/ZnO

Type II

In the type II configuration, the valence and conduction band edge of the core are both lower or higher than the band edges of the shell. An example of a type II is shown in figure X, ZnTe(bandgap:2.26)/CdSe(bandgap:1.74). The lowest energy separation of the electron and the hole will occur when the hole is confined in the ZnTe core valence band and the electron is confined in the CdSe shell conduction band. The emission wavelength will be determined by the energy difference between these occupied states, as shown by the red arrow, which will be at a lower energy than either of the individual bandgaps. The emission wavelength can be significantly red shifted compared to the unpassivated core.

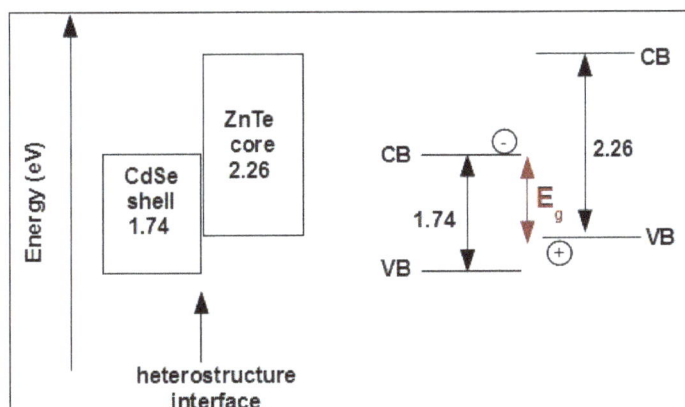

Examples: ZnTe/CdSe, CdTe/CdSe, CdS/ZnSe

Doped Core-shell Semiconductor Nanocrystals

Doping has been shown to strongly affect the optical properties of semiconductor nanocrystals. Impurity concentrations in semiconductor nanocrystals grown using colloidal synthesis, however, are typically lower than in their bulk counterparts. There has been interest in magnetic doping of CSSNCs for applications in magnetic memory and spin-based electronics. Dual-mode optical and magnetic resonance (MR) imaging has been explored by doping the shell of CdSe/ZnS with Mn, which caused the CSSNC to be paramagnetic.

Synthesis

In synthesizing core shell nanoparticles, scientists have studied and found several wet chemical methods, such as chemical precipitation, sol-gel, microemulsion and inverse micelle formation. Those methods have been used to grow core shell chalcogenide nanoparticles with an emphasis on better control of size, shape, and size distribution. To control the growth of nanoparticles with tunable optical properties, supporting matrices such as glasses, zeolites, polymers or fatty acids have been used. In addition, to prepare nanoparticles of sulfides, selenides and tellurides, the Langmuir–Blodgett film technique has been used successfully. In comparison to wet chemical methods, electrochemical synthesis is more desirable, such as the use of aqueous solvents rather than toxic organic solvents, formation of conformal deposits, room-temperature deposition, low cost, and precise control of composition and thickness of semiconductor coating on metal nanoparticles. However, owing to the difficulty of preparing electrically addressable arrays of nanoparticles, the use of electrochemical techniques to produce core-shell nanoparticles was difficult. Recently, Cadmium Sulfide (CdS) and Copper iodide (CuI) was electrochemically grown on a 3-D nanoelectrode array via layer-by-layer depositing of alternating layers of nanoparticles and Polyoxometalate(POM).

Core–shell semiconductor nanocrystals can be grown by using colloidal chemistry methods with an appropriate control of the reaction kinetics. Using this method which results in a relatively high control of size and shape, semiconductor nanostructures could be synthesized in the form of dots, tubes, wires and other forms which show interesting optic and electronic size-dependent properties. Since the synergistic properties resulting from the intimate contact and interaction between the core and shell, CSSNCs can provide novel functions and enhanced properties which are not observed in single nanoparticles.

The size of core materials and the thickness of shell can be controlled during synthesis. For example, in the synthesis of CdSe core nanocrystals, the volume of H_2S gas can determine the size of core nanocrystals. As the volume of H_2S increases, the size of the core decreases. Alternatively, when the reaction solution reaches the desired reaction temperature, rapid cooling can result in smaller core sizes. In addition, the thickness of shell is typically determined by the added amount of shell material during the coating process.

Characterization

An increase in either the core size or shell length results in longer emission wavelengths. The interface between the core and shell can be tailored to passivate relaxation pathways and form radiative states. The size dependence of the band gap in these nanoparticles due to the quantum confinement effect has been utilized to control the photoluminescence color from blue to red by preparing nanoparticles of varying sizes. By manipulating the size or shape of the nanoparticles, the luminescence colors and purity can be controlled. However, the quantum yield and the brightness of luminescence of the CSSNCs is ultimately limited and it cannot be controlled because of the presence of surface traps.

UV-vis absorption spectra, X-ray diffraction (XRD), transmission electron microscopy (TEM) and X-ray photoelectron spectroscopy (XPS) are the techniques typically used to identify and characterize CSSNCs.

Applications

One of the most important properties of core–shell semiconducting nanocrystals (CSSNCs) is that their cores, which are quantum dots, fluoresce, which is important in their biomedical and optical applications. The shells are highly modular, and thus the bulk properties, such as solubility and activity of the CSSNCs can be changed.

Biomedical Applications

The properties desired of CSSNCs when using them for biological applications include high quantum yield, narrow fluorescence emission, broad absorption profile, stability against photobleaching, 20 second fluorescent lifetime, and high brightness. High quantum yields mean that minimal energy will need to be put into the quantum dot to induce fluorescence. A narrow fluorescence emission allows for multiple colors to be imaged at once without color overlap between different types of CSSNCs. Having a broad absorption profile allows multiple CSSNCs to be excited at the same wavelength and thus, multiple CSSNCs could be imaged simultaneously. Having a 20-second fluorescent lifetime allows for time-resolved bioimaging. The utility of CSSNCs is that they can be a complement to organic fluorophores. CSSNCs are less susceptible to photobleaching, but less is known about them compared to organic fluorophores. CSSNCs have 100–1000 times the two-photon fluorescence efficiency as organic dyes, exemplifying their value. In the cases where CSSNCs are used in biological medium, the core is a quantum dot and the shell can be an organic molecule or biological ligands, such as a DNA, that are used for biocompatibility and targeting. The shell can also be an organic molecule to which a biological molecule is later conjugated, furthering the modularity of core–shell structure. The most popular core/shell pair used is CdSe core with ZnS or CdS shell, which improves the quantum yield and protects against photobleaching compared to that of the core material alone. The size of the CSSNC is directly correlated to the color of fluorescence, so being able to control particle size is desirable. However, it is generally unknown how the shell molecules, and salt concentration, pH, and temperature of the media affect the CSSNCs' properties and remains empirical.

In Vitro Cell Labeling

Because multiple colors can be imaged, CSSNCs' ability to be used in cell labeling is of growing importance. However, it can be difficult to get CSSNCs across the cell membrane. This has been achieved via endocytosis (the most common method), direct microinjection, and electroporation, and once in the cell, they become concentrated in the nucleus and can stay there for extended periods of time. Once CSSNCs are inside cells, they remain even after cellular division and can be imaged in both mother and daughter cells. This particular technique was shown using *Xenopus* embryos. Another example of CSSNCs is seen in their tracking ability; when cells are gown on a 2D matrix embedded with CSSNCs, cells uptake the CSSNCs as they move, leaving a trail seen as the absence of CSSNCs. This means that the mobility of cells can be imaged, which is important since the metastatic potential of breast tissue cells has been shown to increase with mobility. Also, it has been shown that five different toxins can be detected using five different CSSNCs simultaneously.

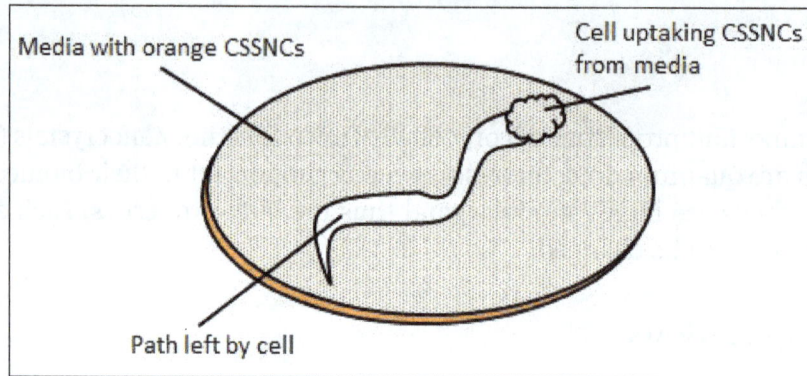

In a move toward environmentally friendlier and less toxic CSSNCs, Si quantum dots with various shells have been developed. Si is 10 times safer than Cd and current work is focused on making Si more water-soluble and biocompatible. In particular, Si quantum dots with poly (acrylic acid) and allylamine shells have been used in cell labeling. Other in vitro uses include flow cyclometry, pathogen detection, and genomic and proteomic detection.

In Vivo and Deep Tissue Imaging

Because CSSNCs emit in the near-infrared region (700–900 nm) of the electromagnetic spectrum, imaging them is not complicated by autofluorescence of tissue, which occurs at higher frequencies (400–600 nm), and scattering effects. This has been used in the mapping of sentinel lymph-nodes in cancer surgery in animals. Lymph nodes 1 cm deep were imaged and the excised nodes with CSSNC accumulation were found to have the highest probability for containing metastatic cells. In addition, CSSNCs have been shown to remain fluorescent in cells in vivo for 4 months. To track and diagnose cancer cells, labeled squamous carminoma cell-line U14 cells were used and fluorescent images could be seen after 6h. CSSNCs conjugated to doxorubicin were also used to target, image, and sense prostate cancer cells that express the prostate-specific membrane antigen protein. Using a cancer-specific antibody conjugated to QDs with polymer shells is the most popular in tumor targeted imaging.

The main disadvantage of using CSSNCs for in vivo imaging is the lack of information about their excretion and toxicity. The typical cores used show DNA damage and toxicity toward liver cells, but using shells seems to diminish this effect. The use of other substances in the core, such as rare-earth elements and Si, are being explored to reduce toxicity. Other disadvantages include limited commercial availability, variability in surface chemistry, nonspecific binding, and instrument limitation.

Optics

The size, shape, and composition of the core–shell structure are related to the bandgap, which in turn is related to its optical properties. Thus, by modulating the size, shape, and material of the core, the optics can be tuned and optimized for use in optical devices and applications such as LEDs, detectors, lasers, phosphors, and photovoltaics.

LEDs

Currently, CSSNC LED efficiency is less than that of organic LEDs. However, studies show that they have potential to accomplish what organic LEDs cannot. CSSNC LEDs constructed using multiple layers of CSSNCs resulted in poor conduction, charge imbalance, low luminescence efficiency, and a large number of pinhole defects. LEDs constructed of one monolayer avoid these problems. An advantage of CSSNC LEDs over organic LEDs is that CSSNC LEDs have narrower emissions, as narrow as 32 nm, than organic LEDs, which range from 50–100 nm. Specifically, the core–shell motif is desirable for use in LEDs because of their electroluminescence and photoluminescence quantum efficiencies and their ability to be processed into devices easily. Current aims for LED displays include developing materials with wavelength emissions of 610–620 nm for red displays, 525–530 nm for green displays, and 460–470 nm for blue displays. This is because these wavelengths maximize the perceived power and they lie outside of the National Television System Committee standard color triangle. CSSNCs have been synthesized that meet these wavelength emissions: (CdSe)ZnS for red emission, (CdS)ZnS for blue emission, and $(Cd_xZn_{1-x}Se)Cd_yZn_{1-y}S$ for the green emission. Using CdSe core and ZnS or CdS/ZnS shells, the maximum luminance values of red, orange, yellow and green LEDs were improved to 9,064, 3,200, 4,470 and 3,700 cd m^{-2}, respectively; electroluminescent efficiency (1.1–2.8 cd A21), and turn-on voltages (3–4 V) were also increased.

Lasers

In CSSNCs with only one exciton, absorption and stimulated emission occur equally and in CSSNCs with more than one exciton, non-radiative Auger recombination occurs, which decays optical gain, an important quality in lasers. However, type II CSSNCs, CdS/ZnSe, were used in optical amplification from stimulated emission of single-exiton states, eliminating Auger recombination. This has the advantage that lasing threshold could be lowered under continuous wave excitation, enhancing the potential of CSSNCs as optical gain media. Type II CSSNCs separate the electrons and holes of the exciton pair, which leads to a strong electric field and thus, reducing absorption losses.

Phosphors

By combining the modularity of CSSNCs and stability of organic polymer, a broad range of colors of phosphors were developed. CdSe core/ZnS shell CSSNCs are used to generate bluish green to red colors, and (CdS)ZnS QDs are used to generate violet to blue colors. By mixing the appropriate amounts of the different sizes of CSSNCs, the entire visible range with narrow emission profiles and high photoluminescence quantum yields can be achieved.

Dye-sensitized Solar Cells

ZnO-TiO$_2$ core-shell nano-structures were synthesized with fast electron transport and high surface area combining the properties of ZnO nanorods and TiO$_2$ nano particles. As ZnO nanorods have fast electron transport and TiO$_2$ nano-particles have high surface area. ZnO-MgO core-shell nanowires were synthesized improving the efficiency of the dye sensitized solar cells by 400% when compared to the ZnO nanowires. MgO shell acts as efficient insulating tunnel preventing recombination.

QUANTUM DOT

Quantum dots are artificial nanostructures that can possess many varied properties, depending on their material and shape. For instance, due to their particular electronic properties they can be used as active materials in single-electron transistors.

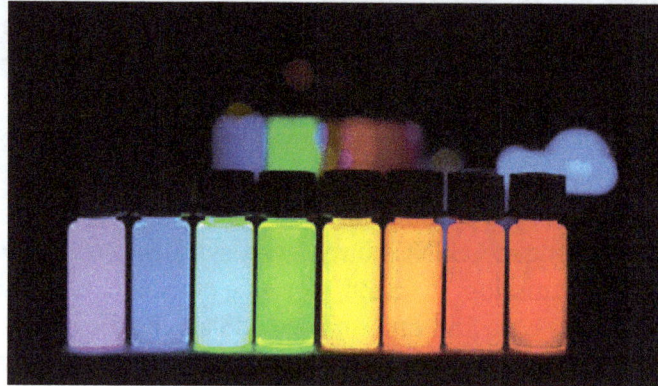

Vials of quantum dots producing vivid colors
from violet to deep red.

The properties of a quantum dot are not only determined by its size but also by its shape, composition, and structure, for instance if it's solid or hollow. A reliable manufacturing technology that makes use of quantum dots' properties – for a wide-ranging number of applications in such areas as catalysis, electronics, photonics, information storage, imaging, medicine, or sensing – needs to be capable of churning out large quantities of nanocrystals where each batch is produced according to the exactly same parameters.

Because certain biological molecules are capable of molecular recognition and self-assembly, nanocrystals could also become an important building block for self-assembled functional nanodevices.

The atom-like energy states of QDs furthermore contribute to special optical properties, such as a particle-size dependent wavelength of fluorescence; an effect which is used in fabricating optical probes for biological and medical imaging.

So far, the use in bioanalytics and biolabeling has found the widest range of applications for colloidal QDs. Though the first generation of quantum dots already pointed out their potential, it took a lot of effort to improve basic properties, in particular colloidal stability in salt-containing solution. Initially, quantum dots have been used in very artificial environments, and these particles would have simply precipitated in 'real' samples, such as blood. These problems have been solved and QDs have found numerous use in real applications.

Quantum dots have found applications in composites, solar cells (Grätzel cells) and fluorescent biological labels (for example to trace a biological molecule) which use both the small particle size and tuneable energy levels.

Advances in chemistry have resulted in the preparation of monolayer-protected, high-quality, monodispersed, crystalline quantum dots as small as 2 nm in diameter, which can be conveniently treated and processed as a typical chemical reagent.

Quantum Dots in Medicine

Quantum dots enable researchers to study cell processes at the level of a single molecule and may significantly improve the diagnosis and treatment of diseases such as cancers. QDs are either used as active sensor elements in high-resolution cellular imaging, where the fluorescence properties of the quantum dots are changed upon reaction with the analyte, or in passive label probes where selective receptor molecules such as antibodies have been conjugated to the surface of the dots.

Quantum dots could revolutionize medicine. Unfortunately, most of them are toxic. Ironically, the existence of heavy metals in QDs such as cadmium, a well-established human toxicant and carcinogen, poses potential dangers especially for future medical application, where qdots are deliberately injected into the body.

As the use of nanomaterials for biomedical applications is increasing, environmental pollution and toxicity have to be addressed, and the development of a non-toxic and biocompatible nanomaterial is becoming an important issue.

Quantum Dots in Photovoltaics

The attractiveness of using quantum dots for making solar cells lies in several advantages over other approaches: They can be manufactured in an energy-saving room-temperature process; they can be made from abundant, inexpensive materials that do not require extensive purification, as silicon does; and they can be applied to a variety of inexpensive and even flexible substrate materials, such as lightweight plastics.

Although using quantum dots as the basis for solar cells is not a new idea, attempts to make photovoltaic devices have not yet achieved sufficiently high efficiency in converting sunlight to power.

A promising route for quantum dot solar cells is a semiconductor ink with the goal of enabling the coating of large areas of solar cell substrates in a single deposition step and thereby eliminating tens of deposition steps necessary with the previous layer-by-layer method.

Graphene Quantum Dots

Graphene, which basically is an unrolled, planar form of a carbon nanotube therefore has become an extremely interesting candidate material for nanoscale electronics. Researchers have shown that it is possible to carve out nanoscale transistors from a single graphene crystal (i.e. graphene quantum dots). Unlike all other known materials, graphene remains highly stable and conductive even when it is cut into devices one nanometer wide.

Graphene quantum dots (GQDs) also show great potential in the fields of photoelectronics, photovoltaics, biosensing, and bioimaging owing to their unique photoluminescence (PL) properties, including excellent biocompatibility, low toxicity, and high stability against photobleaching and photoblinking.

Scientists still are working on finding efficient and universal methods for the synthesis of GQDs with high stability, controllable surface properties, and tunable PL emission wavelength.

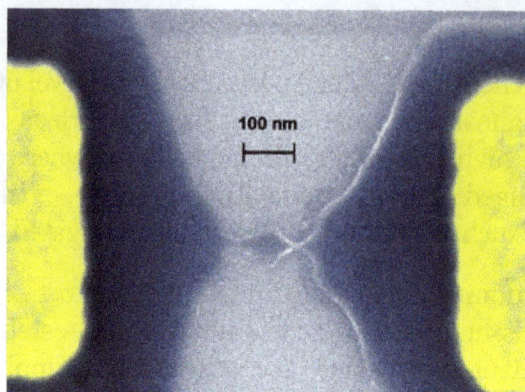

Quantum dot carved from a graphene sheet.

Perovskite Quantum Dots

Luminescent quantum dots (LQDs), which possess high photoluminescence quantum yields, flexible emission color controlling, and solution processibility, are promising for applications in lighting systems (warm white light without UV and infrared irradiation) and high quality displays.

However, the commercialization of LQDs has been held back by the prohibitively high cost of their production. Currently, LQDs are prepared by the HI method, requiring at high temperature and tedious surface treating in order to improve both optical properties and stability.

Although developed only recently, inorganic halide perovskite quantum dot systems have exhibited comparable and even better performances than traditional QDs in many fields. By preparing highly emissive inorganic perovskite quantum dots (IPQDs) at room temperature, IPQDs' superior optical merits could lead to promising applications in lighting and displays.

Quantum Dot TVs and Displays

Quantum dots, because they are both photo-active (photoluminescent) and electro-active (electroluminescent) and have unique physical properties, will be at the core of next-generation displays. Compared to organic luminescent materials used in organic light emitting diodes (OLEDs), QD-based materials have purer colors, longer lifetime, lower manufacturing cost, and lower power consumption. Another key advantage is that, because QDs can be deposited on virtually any substrate, you can expect printable and flexible – even rollable – quantum dot displays of all sizes.

References

- Nanoparticle, science: britannica.com, Retrieved 08 August, 2019

- Gyu-Chul Yi, Chunrui Wang & Won Il Park (2005). "ZnO nanorods: synthesis, characterization and applications". Semiconductor Science and Technology. 20 (4): S22–S34. Bibcode:2005SeScT..20S..22Y. CiteSeerX 10.1.1.453.931. doi:10.1088/0268-1242/20/4/003

- What-is-graphene: nanowerk.com, Retrieved 17 May, 2019

- Brinson, Be; Lassiter, Jb; Levin, Cs; Bardhan, R; Mirin, N; Halas, Nj (November 2008). "Nanoshells Made Easy: Improving Au Layer Growth on Nanoparticle Surfaces". Langmuir. 24 (24): 14166–14171. doi:10.1021/la802049p. PMC 5922771. PMID 19360963

- What-are-quantum-dots: nanowerk.com, Retrieved 01 April, 2019

Nanoelectronics

The utilization of nanotechnology in electronic components is referred to as nanoelectronics. Various components of nanoelectronics are nanoradios, nanomotors, nanoionic devices, nanobatteries, nanogenerators, nanosensors and nanowires. The topics elaborated in this chapter will help in gaining a better perspective about these components of nanoelectronics.

Nanoelectronics is based on the application of nanotechnology in the field of electronics and electronic components. Although the term Nanoelectronics may generally mean all the electronic components, special attention is given in the case of transistors. These transistors have a size lesser than 100 nanometres. Visibly, they are very small that separate studies have to be made for knowing the quantum mechanical properties and inter-atomic design. As a result, though the transistors appear in the nanometre range, they are designed through nanotechnology. Their design is also very much different from the traditional transistors and usually falls in the category of one-dimensional nanotubes/nanowires, hybrid molecular electronics, or advanced molecular electronics.

Basic Concept of Nanoelectronics

Although a nanoelectronic device can be made fully functional, the work load it can do is restricted to its size. The basic principle is that the power of a machine will increase according to the increase in volume, but the amount of friction that the machine's bearings hold will depend on the surface area of the machine.

For the small size of the nanoelectronic device cannot be used for the moving of heavy load like a mechanical device. If such a task is tried, it will fail as the available power will be easily overcome by the frictional forces. So, it is sure that these devices have limitations in real world applications.

Different Approaches to Nanoelectronics

- Nanofabrication:

  This method is used to design arrays or layers of nanoelectronic device to work for a single operation. Nanoelectromechanical systems are also a part of nanofabrication.

- Nanomaterials electronics:

  In Nanoelectronics, the transistors are packed as arrays on to a single chip. Thus they remain in a uniform manner and symmetrical in nature. Thus they are known to have a more speedy movement of electrons in the material. The dielectric constant of the device also increases and the electron or hole characteristics also become symmetrical in nature.

Applications of Nanoelectronics

Some of the devices that have been developed with the help of Nanoelectronics and its future applications are listed below:

- Nanoradio.

- Nanocomputers:

  The conventional computers with a big processor will be replaced with Nanocomputers with nanoprocessors that will have higher performance and speed than the conventional computers. Researchers are performing various experiments on by using nanolithographic methods to design better nanoprocessors. Experiments are also taking place by replacing the CMOS components in conventional processors with nanowires. The FET's in the computers are replaced by carbon nanotubes.

- Energy production:

  The devices using Nanoelectronics technology also includes solar cells that are highly efficient and cheaper than the conventional ones. If such efficient solar energy can be created it would be a revolution to the global energy needs.

Using the technology, researchers are developing a generator for energy production in vivo called bio-nano generators. Basically, the generator is an electrochemical device which is designed in nanoscale size. It works like a fuel cell which generates the power by absorbing the blood glucose in a living body. The glucose will be separated from the body with the help of an enzyme. This enzyme separates the glucose from the electrons and makes them useful for generating power.

The power generated through such a device will be only a few watts as the body itself needs some glucose for its normal functioning. This small power can be used to power up devices placed inside the body like pacemakers or sugar-fed nanorobots.

MOLECULAR SCALE ELECTRONICS

Molecular scale electronics, also called single-molecule electronics, is a branch of nanotechnology that uses single molecules, or nanoscale collections of single molecules, as electronic components. Because single molecules constitute the smallest stable structures imaginable, this miniaturization is the ultimate goal for shrinking electrical circuits.

The field is often termed simply as "molecular electronics", but this term is also used to refer to the distantly related field of conductive polymers and organic electronics, which uses the properties of molecules to affect the bulk properties of a material.

Conventional electronics have traditionally been made from bulk materials. Ever since their invention in 1958, the performance and complexity of integrated circuits has undergone exponential growth, a trend named Moore's law, as feature sizes of the embedded components have shrunk accordingly. As the structures shrink, the sensitivity to deviations increases. In a few technology generations, when the minimum feature sizes reaches 13 nm, the composition of the devices must be controlled to a precision of a few atoms for the devices to work. With bulk methods growing increasingly demanding and costly as they near inherent limits, the idea was born that the components could instead be built up atom by atom in a chemistry lab (bottom up) versus carving them out of bulk material (top down). This is the idea behind molecular electronics, with the ultimate miniaturization being components contained in single molecules.

In single-molecule electronics, the bulk material is replaced by single molecules. Instead of forming structures by removing or applying material after a pattern scaffold, the atoms are put together in a chemistry lab. In this way, billions of billions of copies are made simultaneously (typically more than 10^{20} molecules are made at once) while the composition of molecules are controlled down to the last atom. The molecules used have properties that resemble traditional electronic components such as a wire, transistor or rectifier.

Single-molecule electronics is an emerging field, and entire electronic circuits consisting exclusively of molecular sized compounds are still very far from being realized. However, the unceasing demand for more computing power, along with the inherent limits of lithographic methods as of 2016, make the transition seem unavoidable. Currently, the focus is on discovering molecules with interesting properties and on finding ways to obtain reliable and reproducible contacts between the molecular components and the bulk material of the electrodes.

Molecular electronics operates in the quantum realm of distances less than 100 nanometers. The miniaturization down to single molecules brings the scale down to a regime where quantum mechanics effects are important. In conventional electronic components, electrons can be filled in or drawn out more or less like a continuous flow of electric charge. In contrast, in molecular electronics the transfer of one electron alters the system significantly. For example, when an electron has been transferred from a source electrode to a molecule, the molecule gets charged up, which makes it far harder for the next electron to transfer. The significant amount of energy due to charging must be accounted for when making calculations about the electronic properties of the setup, and is highly sensitive to distances to conducting surfaces nearby.

The theory of single-molecule devices is especially interesting since the system under consideration is an open quantum system in nonequilibrium (driven by voltage). In the low bias voltage regime, the nonequilibrium nature of the molecular junction can be ignored, and the current-voltage traits of the device can be calculated using the equilibrium electronic structure of the system. However, in stronger bias regimes a more sophisticated treatment is required, as there is no longer a variational principle. In the elastic tunneling case (where the passing electron does not exchange energy with the system), the formalism of Rolf Landauer can be used to calculate the transmission through the system as a function of bias voltage, and hence the current. In inelastic tunneling, an elegant formalism based on the non-equilibrium Green's functions of Leo Kadanoff and Gordon Baym, and independently by Leonid Keldysh was advanced by Ned Wingreen and Yigal Meir. This Meir-Wingreen formulation has been used to great success in the molecular electronics community to examine the more difficult and interesting cases where the transient electron exchanges energy with the molecular system (for example through electron-phonon coupling or electronic excitations).

Further, connecting single molecules reliably to a larger scale circuit has proven a great challenge, and constitutes a significant hindrance to commercialization.

Examples:

Common for molecules used in molecular electronics is that the structures contain many alternating double and single bonds. This is done because such patterns delocalize the molecular orbitals, making it possible for electrons to move freely over the conjugated area.

Wires

The sole purpose of molecular wires is to electrically connect different parts of a molecular electrical circuit. As the assembly of these and their connection to a macroscopic circuit is still not mastered, the focus of research in single-molecule electronics is primarily on the functionalized molecules: molecular wires are characterized by containing no functional groups and are hence composed of plain repetitions of a conjugated building block. Among these are the carbon nanotubes that are quite large compared to the other suggestions but have shown very promising electrical properties.

The main problem with the molecular wires is to obtain good electrical contact with the electrodes so that electrons can move freely in and out of the wire.

Transistors

Single-molecule transistors are fundamentally different from the ones known from bulk electronics. The gate in a conventional (field-effect) transistor determines the conductance between the source and drain electrode by controlling the density of charge carriers between them, whereas the gate in a single-molecule transistor controls the possibility of a single electron to jump on and off the molecule by modifying the energy of the molecular orbitals. One of the effects of this difference is that the single-molecule transistor is almost binary: it is either *on* or *off*. This opposes its bulk counterparts, which have quadratic responses to gate voltage.

It is the quantization of charge into electrons that is responsible for the markedly different behavior compared to bulk electronics. Because of the size of a single molecule, the charging due to a single electron is significant and provides means to turn a transistor *on* or *off*. For this to work, the electronic orbitals on the transistor molecule cannot be too well integrated with the orbitals on the electrodes. If they are, an electron cannot be said to be located on the molecule or the electrodes and the molecule will function as a wire.

A popular group of molecules, that can work as the semiconducting channel material in a molecular transistor, is the oligopolyphenylenevinylenes (OPVs) that works by the Coulomb blockade mechanism when placed between the source and drain electrode in an appropriate way. Fullerenes work by the same mechanism and have also been commonly used.

Semiconducting carbon nanotubes have also been demonstrated to work as channel material but although molecular, these molecules are sufficiently large to behave almost as bulk semiconductors.

The size of the molecules, and the low temperature of the measurements being conducted, makes the quantum mechanical states well defined. Thus, it is being researched if the quantum mechanical properties can be used for more advanced purposes than simple transistors (e.g. spintronics).

Physicists at the University of Arizona, in collaboration with chemists from the University of Madrid, have designed a single-molecule transistor using a ring-shaped molecule similar to benzene. Physicists at Canada's National Institute for Nanotechnology have designed a single-molecule transistor using styrene. Both groups expect (the designs were experimentally unverified as of June 2005) their respective devices to function at room temperature, and to be controlled by a single electron.

Rectifiers (Diodes)

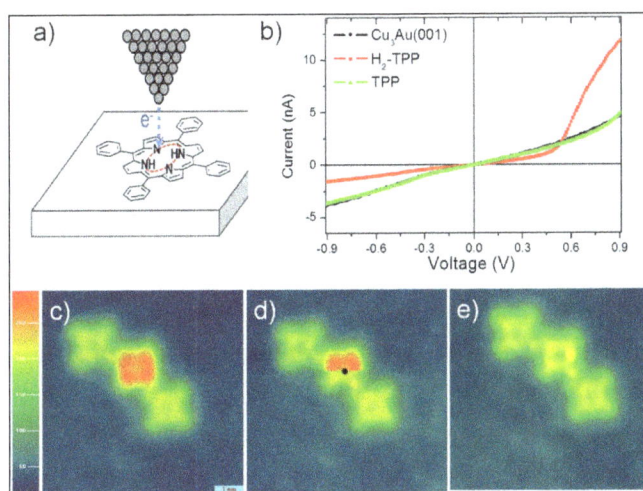

In figure, hydrogen can be removed from individual tetraphenylporphyrin (H_2TPP) molecules by applying excess voltage to the tip of a scanning tunneling microscope (STAM, a); this removal alters the current-voltage (I-V) curves of TPP molecules, measured using the same STM tip, from diode-like (red curve in b) to resistor-like (green curve). Figure (c) shows a row of TPP, H_2TPP and TPP molecules. While scanning image (d), excess voltage was applied to H_2TPP at the black dot, which instantly removed hydrogen, as shown in the bottom part of (d) and in the re-scan image (e). Such manipulations can be used in single-molecule electronics.

Molecular rectifiers are mimics of their bulk counterparts and have an asymmetric construction so that the molecule can accept electrons in one end but not the other. The molecules have an electron donor (D) in one end and an electron acceptor (A) in the other. This way, the unstable state $D^+ - A^-$ will be more readily made than $D^- - A^+$. The result is that an electric current can be drawn through the molecule if the electrons are added through the acceptor end, but less easily if the reverse is attempted.

Methods

One of the biggest problems with measuring on single molecules is to establish reproducible electrical contact with only one molecule and doing so without shortcutting the electrodes. Because the current photolithographic technology is unable to produce electrode gaps small enough to contact both ends of the molecules tested (on the order of nanometers), alternative strategies are applied.

Molecular Gaps

One way to produce electrodes with a molecular sized gap between them is break junctions, in which a thin electrode is stretched until it breaks. Another is electromigration. Here a current is led through a thin wire until it melts and the atoms migrate to produce the gap. Further, the reach of conventional photolithography can be enhanced by chemically etching or depositing metal on the electrodes.

Probably the easiest way to conduct measurements on several molecules is to use the tip of a scanning tunneling microscope (STM) to contact molecules adhered at the other end to a metal substrate.

Anchoring

A popular way to anchor molecules to the electrodes is to make use of sulfur's high chemical affinity to gold. In these setups, the molecules are synthesized so that sulfur atoms are placed strategically to function as crocodile clips connecting the molecules to the gold electrodes. Though useful, the anchoring is non-specific and thus anchors the molecules randomly to all gold surfaces. Further, the contact resistance is highly dependent on the precise atomic geometry around the site of anchoring and thereby inherently compromises the reproducibility of the connection.

To circumvent the latter issue, experiments has shown that fullerenes could be a good candidate for use instead of sulfur because of the large conjugated π-system that can electrically contact many more atoms at once than one atom of sulfur.

Fullerene Nanoelectronics

In polymers, classical organic molecules are composed of both carbon and hydrogen (and sometimes additional compounds such as nitrogen, chlorine or sulphur). They are obtained from petrol and can often be synthesized in large amounts. Most of these molecules are insulating when their length exceeds a few nanometers. However, naturally occurring carbon is conducting, especially graphite recovered from coal or encountered otherwise. From a theoretical viewpoint, graphite is a semi-metal, a category in between metals and semi-conductors. It has a layered structure, each sheet being one atom thick. Between each sheet, the interactions are weak enough to allow an easy manual cleavage.

Tailoring the graphite sheet to obtain well defined nanometer-sized objects remains a challenge. However, by the close of the twentieth century, chemists were exploring methods to fabricate extremely small graphitic objects that could be considered single molecules. After studying the interstellar conditions under which carbon is known to form clusters, Richard Smalley's group (Rice University, Texas) set up an experiment in which graphite was vaporized via laser irradiation. Mass spectrometry revealed that clusters containing specific *magic numbers* of atoms were stable, especially those clusters of 60 atoms. Harry Kroto, an English chemist who assisted in the experiment, suggested a possible geometry for these clusters – atoms covalently bound with the exact symmetry of a soccer ball. Coined buckminsterfullerenes, buckyballs, or C_{60}, the clusters retained some properties of graphite, such as conductivity. These objects were rapidly envisioned as possible building blocks for molecular electronics.

Problems

Artifacts

When trying to measure electronic traits of molecules, artificial phenomena can occur that can be hard to distinguish from truly molecular behavior. Before they were discovered, these artifacts have mistakenly been published as being features pertaining to the molecules in question.

Applying a voltage drop on the order of volts across a nanometer sized junction results in a very strong electrical field. The field can cause metal atoms to migrate and eventually close the gap by a thin filament, which can be broken again when carrying a current. The two levels of conductance imitate molecular switching between a conductive and an isolating state of a molecule.

Another encountered artifact is when the electrodes undergo chemical reactions due to the high field strength in the gap. When the voltage bias is reversed, the reaction will cause hysteresis in the measurements that can be interpreted as being of molecular origin.

A metallic grain between the electrodes can act as a single electron transistor by the mechanism described above, thus resembling the traits of a molecular transistor. This artifact is especially common with nanogaps produced by the electromigration method.

Commercialization

One of the biggest hindrances for single-molecule electronics to be commercially exploited is the lack of methods to connect a molecular sized circuit to bulk electrodes in a way that gives reproducible results. At the current state, the difficulty of connecting single molecules vastly outweighs any possible performance increase that could be gained from such shrinkage. The difficulties grow worse if the molecules are to have a certain spatial orientation and have multiple poles to connect.

Also problematic is that some measurements on single molecules are carried out in cryogenic temperatures (near absolute zero), which is very energy consuming. This is done to reduce signal noise enough to measure the faint currents of single molecules.

BALLISTIC CONDUCTION

Ballistic conduction (ballistic transport) is the transport of charge carriers (usually electrons) in a medium, having negligible electrical resistivity caused by scattering. Without scattering, electrons simply obey Newton's second law of motion at non-relativistic speeds.

In general, the resistivity of a material exists because an electron, while moving inside a medium, is scattered by impurities, defects, thermal fluctuations of ions in a crystalline solid, or, generally, by any freely-moving atom/molecule composing a gas or liquid.

For a given particle, a mean free path can be described as being the average length that the electron can travel freely, i.e., before a collision, which could change its momentum. The mean free path can be increased by reducing the number of impurities in a crystal or by lowering its temperature.

Ballistic transport is observed when the mean free path of the electron is (much) longer than the dimension of the medium through which the electron travels. The electron alters its motion only upon collision with the *walls*. In the case of a wire suspended in air/vacuum the surface of the wire plays the role of the *box* reflecting the electrons and preventing them from exiting toward the empty space/open air. This is because there is an energy to be paid to extract the electron from the medium (work function).

For example, ballistic transport can be observed in a metal nanowire: this is simply because the wire is of the size of a nanometer (10^{-9} meters) and the mean free path can be longer than that in a metal.

Ballistic conduction is the unimpeded flow of charge, or energy-carrying particles, over relatively long distances in a material. Normally, transport of electrons (or holes) is dominated by scattering events, which relax the carrier momentum in an effort to bring the conducting material to equilibrium. Thus, ballistic transport in a material is determined by how ballistically conductive that material is. Ballistic conduction differs from superconductivity due to the absence of the Meissner effect in the material. A ballistic conductor would stop conducting if the driving force is turned off, whereas in a superconductor current would continue to flow after the driving supply is disconnected.

Ballistic conduction is typically observed in quasi-1D structures, such as carbon nanotubes or silicon nanowires, because of extreme size quantization effects in these materials. Ballistic conduction is not limited to electrons (or holes) but can also apply to phonons. It is theoretically possible for ballistic conduction to be extended to other quasi-particles, but this has not been experimentally verified.

Scattering Mechanisms

In general, carriers will exhibit ballistic conduction when $L \leq \lambda_{MFP}$ where L is the length of the active part of the device (i.e., a channel in a MOSFET). λ_{MFP} is the mean free path for the carrier which can be given by Matthiessen's Rule, written here for electrons:

$$\frac{1}{\lambda_{MFP}} = \frac{1}{\lambda_{el\text{-}el}} + \frac{1}{\lambda_{ap}} + \frac{1}{\lambda_{op,ems}} + \frac{1}{\lambda_{op,abs}} + \frac{1}{\lambda_{impurity}} + \frac{1}{\lambda_{defect}} + \frac{1}{\lambda_{boundary}}$$

where,

- $\lambda_{el\text{-}el}$ is the electron-electron scattering length,

- λ_{ap} is the acoustic phonon (emission and absorption) scattering length,

- $\lambda_{op,ems}$ is the optical phonon emission scattering length,

- $\lambda_{op,abs}$ is the optical phonon absorption scattering length,

- $\lambda_{impurity}$ is the electron-impurity scattering length,

- λ_{defect} is the electron-defect scattering length,

- and $\lambda_{\text{boundary}}$ is the electron scattering length with the boundary.

In terms of scattering mechanisms, optical phonon emission normally dominates, depending on the material and transport conditions. There are also other scattering mechanisms which apply to different carriers that are not considered here (e.g. remote interface phonon scattering, Umklapp scattering). To get these characteristic scattering rates, one would need to derive a Hamiltonian and solve Fermi's golden rule for the system in question.

Landauer-Büttiker Formalism

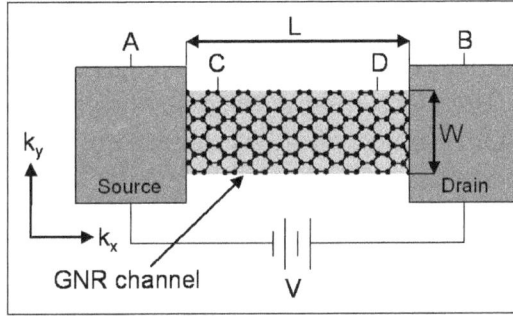

A graphene nanoribbon field-effect transistor (GNR-FET). Here contacts A and B are at two different Fermi levels E_{F_A} and E_{F_B}.

In 1957, Rolf Landauer proposed that conduction in a 1D system could be viewed as a transmission problem. For the 1D graphene nanoribbon field effect transistor (GNR-FET) on the right (where the channel is assumed to be ballistic), the current from A to B, given by the Boltzmann transport equation, is,

$$I_{AB} = \frac{g_s e}{h} \int_{E_{F_B}}^{E_{F_A}} M(E) f'(E) T(E) dE,$$

where, g_s=2, due to spin degeneracy, e is the electron charge, h is the Planck's constant, E_{F_A} and E_{F_B} are the Fermi levels of A and B, $M(E)$ is the number of propagating modes in the channel, $f'(E)$ is the deviation from the equilibrium electron distribution (perturbation), and $T(E)$ is the transmission probability (T=1 for ballistic). Based on the definition of conductance,

$$G = \frac{I}{V},$$

and the voltage separation between the Fermi levels is approximately $eV = E_{F_A} - E_{F_B}$, it follows that,

$$G = G_0 M T, \text{ with } G_0 = \frac{2e^2}{h}$$

where, M is the number of modes in the transmission channel and spin is included. G_0 is known as the conductance quantum. The contacts have a multiplicity of modes due to their larger size in comparison to the channel. Conversely, the quantum confinement in the 1D GNR channel constricts the number of modes to carrier degeneracy and restrictions from the energy dispersion relationship and the Brillouin zone. For example, electrons in carbon nanotubes have two intervalley modes and two spin modes. Since the contacts and the GNR channel are connected by leads, the transmission probability is smaller at contacts A and B,

$$T \approx \frac{M}{M_{\text{contact}}}.$$

Thus the quantum conductance is approximately the same if measured at A and B or C and D.

Landauer-Büttiker formalism holds as long as the carriers are coherent (which means the length of the active channel is less than the phase-breaking mean free path) and the transmission functions can be calculated from Schrödinger's equation or approximated by semiclassical approximations, like the WKB approximation. Therefore, even in the case of a perfect ballistic transport, there is a fundamental ballistic conductance which saturates the current of the device with a resistance of approximately 12.9 kΩ per mode (spin degeneracy included). There is, however, a generalization of the Landauer-Büttiker formalism of transport applicable to time-dependent problems in the presence of dissipation.

Importance

Ballistic conduction enables use of quantum mechanical properties of electron wave functions. Ballistic transport is coherent in wave mechanics terms. Phenomena like double-slit interference, spatial resonance (and other optical or microwave-like effects) could be exploited in electronic systems at nanoscale in systems including nanowires and nanotubes.

The widely encountered phenomenon of electrical contact resistance or ECR, arises as an electric current flowing through a rough interface is restricted to a limited number of contact spots. The size and distribution of these contact spots is governed by the topological structures of the contacting surfaces forming the electrical contact. In particular, for surfaces with high fractal dimension contact spots may be very small. In such cases, when the radius of the contact spot is smaller than the mean free path of electrons λ, the resistance is dominated by the Sharvin mechanism, in which electrons travel ballistically through these micro-contacts with resistance that can be described by the following:

$$R_{\text{S}} = \frac{\lambda(\rho_1 + \rho_2)}{2a}.$$

This term, where ρ_1 and ρ_2 correspond to the specific resistivity of the two contacting surfaces, is known as Sharvin resistance. Electrical contacts resulting in ballistic electron conduction are known as *Sharvin Contacts*. When the radius of a contact spot is larger than the mean free path of electrons, the contact resistance can be treated classically.

Optical Analogies

A comparison with light provides an analogy between ballistic and non-ballistic conduction.

Ballistic electrons behave like light in a waveguide or a high-quality optical assembly. Non-ballistic electrons behave like light diffused in milk or reflected off a white wall or a piece of paper.

Electrons can be scattered several ways in a conductor. Electrons have several properties: wavelength (energy), direction, phase, and spin orientation. Different materials have different scattering probabilities which cause different incoherence rates (stochasticity). Some kinds of scattering can only cause a change in electron direction, others can cause energy loss.

Consider a coherent source of electrons connected to a conductor. Over a limited distance, the electron wave function will remain coherent. You still can deterministically predict its behavior (and use it for computation theoretically). After some greater distance, scattering causes each electron to have a slightly different phase and direction. But there is still almost no energy loss. Like monochromatic light passing through milk, electrons undergo elastic interactions. Information about the state of the electrons at the input is then lost. Transport becomes statistical and stochastic. From the resistance point of view, stochastic (not oriented) movement of electrons is useless even if they carry the same energy – they move thermally. If the electrons undergo inelastic interactions too, they lose energy and the result is a second mechanism of resistance. Electrons which undergo inelastic interaction are then similar to non-monochromatic light.

For correct usage of this analogy consideration of several facts is needed:

1. Photons are bosons and electrons are fermions;

2. There is coulombic repulsion between electrons thus this analogy is good only for single-electron conduction because electron processes are strongly nonlinear and dependent on other electrons;

3. It is more likely that an electron would lose more energy than a photon would, because of the electron's non-zero rest mass;

4. Electron interactions with the environment, each other, and other particles are generally stronger than interactions with and between photons.

Examples

As mentioned, nanostructures such as carbon nanotubes or graphene nanoribbons are often considered ballistic, but these devices only very closely resemble ballistic conduction. Their ballisticity is nearly 0.9 at room temperature.

Carbon Nanotubes and Graphene Nanoribbon

The dominant scattering mechanism at room temperature is that of electrons emitting optical phonons. If electrons don't scatter with enough phonons (for example if the scattering rate is low), the mean free path tends to be very long ($\lambda_{MFP} \approx 1\mu m$). So a nanotube or graphene nanoribbon could be a good ballistic conductor if the electrons in transit don't scatter with too many phonons and if the device is about 100 nm long. Such a transport regime has been found to depend on the nanoribbon edge structure and the electron energy.

Silicon Nanowires

It is often incorrectly thought that Si nanowires are quantum confined ballistic conductors. There are major differences between carbon nanotubes (which are hollow) and Si nanowires (which are solid). Nanowires are about 20–50 nm in diameter and are 3D solid while carbon nanotubes have diameters around the wavelength of the electrons (2–3 nm) and are essentially 1D conductors. However it is still possible to observe ballistic conduction in Si nanowires at very low temperatures (2–3 K).

Isotopically Enriched Diamond

Isotopically pure diamond can have a significantly higher thermal conductivity.

Ballistic Thermal Transport

Heat conduction can experience ballistic thermal transport when heating size is larger than phonon mean free paths. Ballistic thermal transport has been observed in multiple materials systems.

COULOMB BLOCKADE

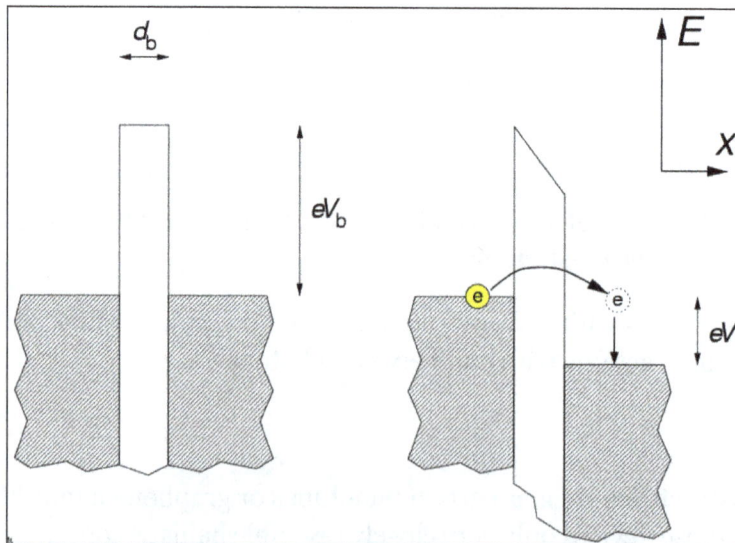

Schematic representation (similar to band diagram)
of an electron tunnelling through a barrier.

A Coulomb blockade (CB), named after Charles-Augustin de Coulomb's electrical force, is the decrease in electrical conductance at small bias voltages of a small electronic device comprising at least one low-capacitance tunnel junction. Because of the CB, the conductance of a device may not be constant at low bias voltages, but disappear for biases under a certain threshold, i.e. no current flows.

Coulomb blockade can be observed by making a device very small, like a quantum dot. When the device is small enough, electrons inside the device will create a strong Coulomb repulsion preventing other electrons to flow. Thus, the device will no longer follow Ohm's law and the current-voltage relation of the Coulomb blockade looks like a staircase.

Even though the Coulomb blockade can be used to demonstrate the quantization of the electric charge, it remains a classical effect and its main description does not require quantum mechanics. However, when few electrons are involved and an external static magnetic field is applied, Coulomb blockade provides the ground for a spin blockade (like Pauli spin blockade) and valley blockade, which include quantum mechanical effects due to spin and orbital interactions respectively between the electrons.

The devices can be composed of either metallic or superconducting electrodes. If the electrodes are superconducting, Cooper pairs (with a charge of minus two elementary charges $-2e$) carry the current. In the case that the electrodes are metallic or *normal-conducting*, i.e. neither superconducting nor semiconducting, electrons (with a charge of $-e$) carry the current.

In a Tunnel Junction

The tunnel junction is, in its simplest form, a thin insulating barrier between two conducting electrodes. According to the laws of classical electrodynamics, no current can flow through an insulating barrier. According to the laws of quantum mechanics, however, there is a nonvanishing (larger than zero) probability for an electron on one side of the barrier to reach the other side. When a bias voltage is applied, this means that there will be a current, and, neglecting additional effects, the tunnelling current will be proportional to the bias voltage. In electrical terms, the tunnel junction behaves as a resistor with a constant resistance, also known as an ohmic resistor. The resistance depends exponentially on the barrier thickness. Typically, the barrier thickness is on the order of one to several nanometers.

An arrangement of two conductors with an insulating layer in between not only has a resistance, but also a finite capacitance. The insulator is also called dielectric in this context, the tunnel junction behaves as a capacitor.

Due to the discreteness of electrical charge, current through a tunnel junction is a series of events in which exactly one electron passes (*tunnels*) through the tunnel barrier (we neglect cotunneling, in which two electrons tunnel simultaneously). The tunnel junction capacitor is charged with one elementary charge by the tunnelling electron, causing a voltage build up $U = e/C$, where C is the capacitance of the junction. If the capacitance is very small, the voltage build up can be large enough to prevent another electron from tunnelling. The electric current is then suppressed at low bias voltages and the resistance of the device is no longer constant. The increase of the differential resistance around zero bias is called the Coulomb blockade.

Observation

In order for the Coulomb blockade to be observable, the temperature has to be low enough so that the characteristic charging energy (the energy that is required to charge the junction with one elementary charge) is larger than the thermal energy of the charge carriers. In the past, for capacitances above 1 femtofarad (10^{-15} farad), this implied that the temperature has to be below about 1 kelvin. This temperature range is routinely reached for example by 3He refrigerators. Thanks to small sized quantum dots of only few nanometers, Coulomb blockade has been observed next above liquid helium temperature, up to room temperature.

To make a tunnel junction in plate condenser geometry with a capacitance of 1 femtofarad, using an oxide layer of electric permittivity 10 and thickness one nanometer, one has to create electrodes with dimensions of approximately 100 by 100 nanometers. This range of dimensions is routinely reached for example by electron beam lithography and appropriate pattern transfer technologies, like the Niemeyer–Dolan technique, also known as shadow evaporation technique. The integration of quantum dot fabrication with standard industrial technology has been achieved for silicon. CMOS process for obtaining massive production of single electron quantum dot transistors with channel size down to 20 nm x 20 nm has been implemented.

Single-electron Transistor

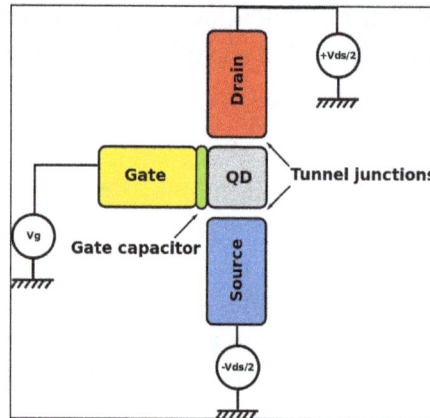

Schematic of a single-electron transistor.

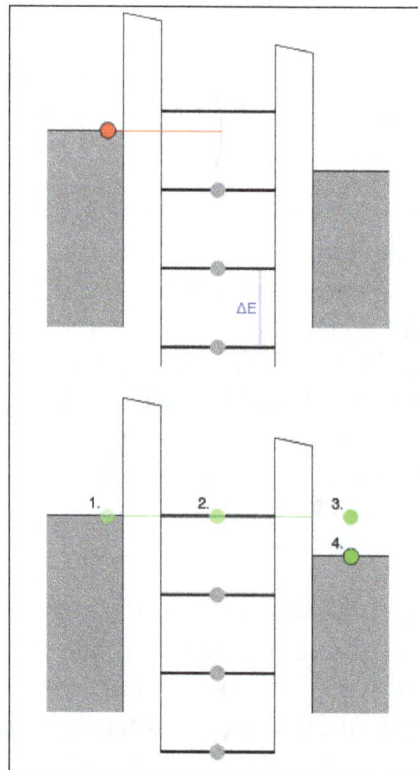

Left to right: energy levels of source, island and drain in a single-electron transistor for the blocking state (upper part) and transmitting state (lower part).

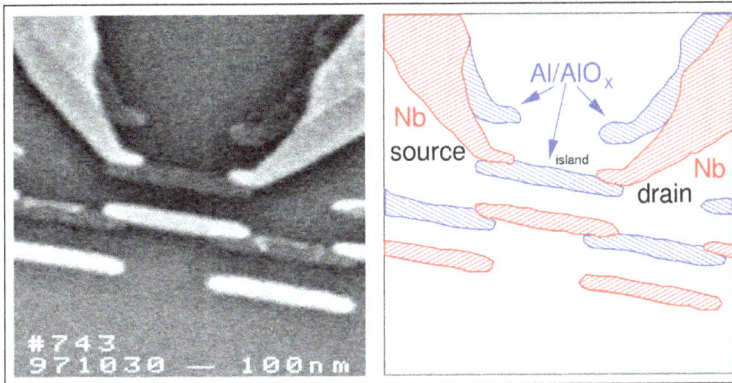

Single-electron transistor with niobium leads and aluminium island.

The simplest device in which the effect of Coulomb blockade can be observed is the so-called single-electron transistor. It consists of two electrodes known as the *drain* and the *source*, connected through tunnel junctions to one common electrode with a low self-capacitance, known as the *island*. The electrical potential of the island can be tuned by a third electrode, known as the *gate*, which is capacitively coupled to the island.

In the blocking state no accessible energy levels are within tunneling range of an electron (in red) on the source contact. All energy levels on the island electrode with lower energies are occupied.

When a positive voltage is applied to the gate electrode the energy levels of the island electrode are lowered. The electron green can tunnel onto the island, occupying a previously vacant energy level. From there it can tunnel onto the drain electrode, where it inelastically scatters and reaches the drain electrode Fermi level.

The energy levels of the island electrode are evenly spaced with a separation of ΔE This gives rise to a self-capacitance C of the island, defined as:

$$C = \frac{e^2}{\Delta E}$$

To achieve the Coulomb blockade, three criteria have to be met:

1. The bias voltage must be lower than the elementary charge divided by the self-capacitance of the island: $V_{\text{bias}} < \frac{e}{C}$;

2. The thermal energy in the source contact plus the thermal energy in the island, i.e. $k_{\text{B}}T$ must be below the charging energy: $k_{\text{B}}T < \frac{e^2}{2C}$ or else the electron will be able to pass the QD via thermal excitation;

3. The tunneling resistance, R_t should be greater than $\frac{h}{e^2}$ which is derived from Heisenberg's uncertainty principle.

Coulomb Blockade Thermometer

A typical Coulomb blockade thermometer (CBT) is made from an array of metallic islands, connected to each other through a thin insulating layer. A tunnel junction forms between the islands, and as voltage is applied, electrons may tunnel across this junction. The tunneling rates and hence the conductance vary according to the charging energy of the islands as well as the thermal energy of the system.

Coulomb blockade thermometer is a primary thermometer based on electric conductance characteristics of tunnel junction arrays. The parameter $V_{1/2}$=5.439$Nk_B T/e$, the full width at half minimum of the measured differential conductance dip over an array of N junctions together with the physical constants provide the absolute temperature.

Ionic Coulomb Blockade

Ionic Coulomb blockade (ICB) is an electrostatic phenomenon that appears in ionic transport through mesoscopic electro-diffusive systems (artificial nanopores and biological ion channels) and manifests itself as oscillatory dependences of the conductance on the fixed charge Q_f in the pore (or on the external voltage V, or on the bulk concentration c_b).

ICB represents an ion-related counterpart of the better-known electronic Coulomb blockade (ECB) that is observed in quantum dots. Both ICB and ECB arise from quantisation of the electric charge and from an electrostatic exclusion principle and they share in common a number of effects and underlying physical mechanisms. ICB provides some specific effects related to the existence of ions of different charge $q = ze$ (different in both sign and value) where integer z is ion valence and e is the elementary charge, in contrast to the single-valence electrons of ECB ($z = -1$).

ICB effects appear in tiny pores whose self-capacitance C_s is so small that the charging energy of a single ion $\Delta E = z^2 e^2 / (2C_s)$ becomes large compared to the thermal energy per particle ($\Delta E \gg k_B T$). In such cases there is strong quantisation of the energy spectrum inside the pore, and the system may either be "blockaded" against the transportation of ions or, in the opposite extreme, it may show resonant barrier-less conduction, depending on the free energy bias coming from Q_f, V, or $\log c_b$.

The ICB model claims that Q_f is a primary determinant of conduction and selectivity for particular ions, and the predicted oscillations in conductance and an associated Coulomb staircase of channel occupancy vs Q_f are expected to be strong effects in the cases of divalent ions ($z = 2$) or trivalent ions ($z = 3$).

Some effects, now recognised as belonging to ICB, were discovered and considered earlier in precursor papers on electrostatics-governed conduction mechanisms in channels and nanopores.

The manifestations of ICB have been observed in water-filled sub-nanometre pores through a 2D MoS_2 monolayer, revealed by Brownian dynamics (BD) simulations of calcium conductance bands in narrow channels, and account for a diversity of effects seen in biological ion channels. ICB predictions have also been confirmed by a mutation study of divalent blockade in the NaChBac bacterial channel.

Model

Generic Electrostatic Model of Channel/Nanopore

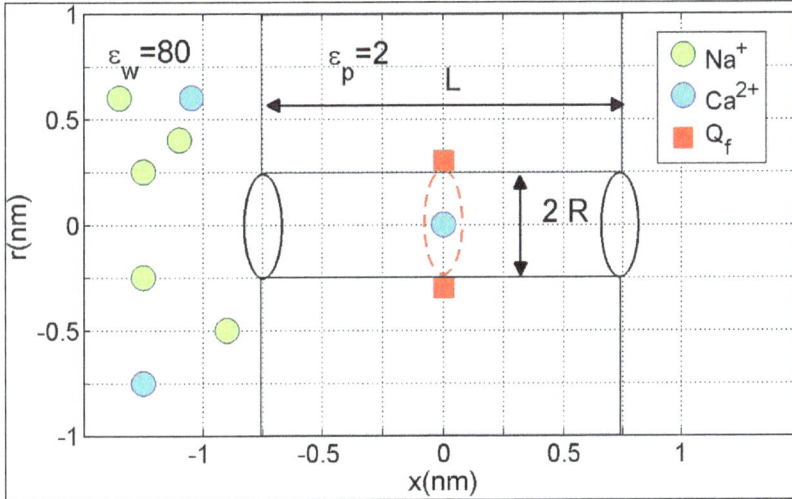

Generic electrostatic and Brownian dynamics model of a channel or nanopore.

ICB effects may be derived on the basis of a simplified electrostatics/Brownian dynamics model of a nanopore or of the selectivity filter of an ion channel. The model represents the channel/pore as a charged hole through a water-filled protein hub embedded in the membrane. Its fixed charge Q_f is considered as a uniform, centrally placed, rigid ring. The channel is assumed to have geometrical parameters length $L \approx 1\,\text{nm}$ and radius $R \approx 0.3 - 0.5\,\text{nm}$, allowing for the single-file movement of partially hydrated ions.

The model represents the water and protein as continuous media with dielectric constants $\varepsilon_w = 80$ and $\varepsilon_p = 2 - 10$ respectively. The mobile ions are described as discrete entities with valence z and of radius R_{ion}, moving stochastically through the pore, governed by the self-consistently coupled Poisson's electrostatic equation and Langevin stochastic equation.

The model is applicable to both cationic and anionic biological ion channels and to artificial nanopores.

Electrostatics

The mobile ion is assumed to be partially hydrated (typically retaining its first hydration shell) and carrying charge $q = ze$ where e is the elementary charge (e.g. the Ca^{2+} ion with $z = 2$). The model allows one to derive the pore and ion parameters satisfying the barrier-less permeation conditions, and to do so from basic electrostatics taking account of charge quantisation.

The potential energy E_n of a channel/pore containing n ions can be decomposed into electrostatic energy E_n^{ES}, dehydration energy, E_n^{DH} and ion-ion local interaction energy E_n^{INT}:

$$E_n = E_n^{ES} + E_n^{DH} + E_n^{INT} ...(E_n \text{ Decomposition})$$

The basic ICB model makes the simplifying approximation that $E_n = E_n^{ES}$, whence:

$$Q_n = zen + Q_f \text{ (Excess charge)}$$

$$E_n = \frac{Q_n^2}{2C_s} \text{ (Electrostatic energy)}$$

$$C_s = 4\pi\epsilon_0\epsilon_w \frac{R^2}{L} \text{ (Self-capacitance)}$$

where, Q_n is the net charge of the pore when it contains n identical ions of valence z, the sign of the moving ions being opposite to that of the Q_f, C_s represents the electrostatic self-capacitance of the pore, and ϵ_0 is the electric permittivity of the vacuum.

Resonant Barrier-less Conduction

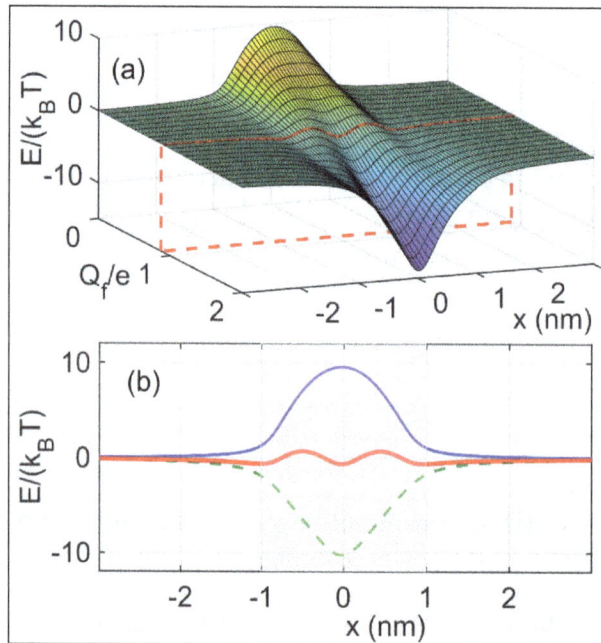

Resonant barrier-less conduction of Ca^{2+} ions, with energies E plotted vertically. (a) Plot of μ_{ex} as a function of fixed charge Q_f/e and position x in the channel. At the "resonant" value of $Q_f/e = 1$ the transition is almost barrier-less (red cross-section). (b) Plots of ΔE (blue curve) and E_{AFF} (dashed-green) and their sum μ_{ex} (red) against x for $Q_f/e = 1$, showing that barrier-less conduction originates in a near-cancellation between ΔE and E_{AFF}.

Thermodynamics and statistical mechanics describe systems that have variable numbers of particles via the chemical potential μ, defined as Gibbs free energy G per particle:

$$G_n = E_n - TS_n \text{ (Gibbs free energy)}$$

$$\mu_n = G_{n+1} - G_n \text{ (Chemical potential)}$$

where, G_n is the Gibbs free energy for the system of n particles. In thermal and particle equilibrium with bulk reservoirs, the entire system has a common value of chemical potential $\mu = \mu_F$ (the Fermi level in other contexts). The free energy needed for the entry of a new ion to the channel is defined by the excess chemical potential $\mu_{ex} = \mu_n - \mu_F$ which (ignoring an entropy term) can be written as:

$$\mu_{ex} = E_{n+1} - E_n = \Delta E + E_{AFF} \quad \text{(Coulomb gap)}$$

$$\Delta E = \frac{z^2 e^2}{2C_s}; \qquad\qquad \text{(Charging energy)}$$

$$E_{AFF} = \frac{ze}{C_s}(zen + Q_f) \qquad \text{(Affinity energy)}$$

where, ΔE is the charging energy (self-energy barrier) of an incoming ion and E_{AFF} is its affinity (i.e. energy of attraction to the binding site Q_f). The difference in energy between ΔE and E_{AFF} defines the ionic energy level separation (Coulomb gap) and gives rise to most of the observed ICB effects.

In selective ion channels, the favoured ionic species passes through the channel almost at the rate of free diffusion, despite the strong affinity to the binding site. This conductivity-selectivty paradox has been explained as being a consequaence of selective barrier-less conduction. In the ICB model, this occurs when ΔE is almost exactly balanced by E_{AFF} ($\mu_{ex} \approx 0$), which happens for a particular value of Q_f . This resonant value of Q_f depends on the ionic properties z and R_{ion} (implicitly, via the R_{ion} -dependent dehydration energy).

Oscillations of Conductance

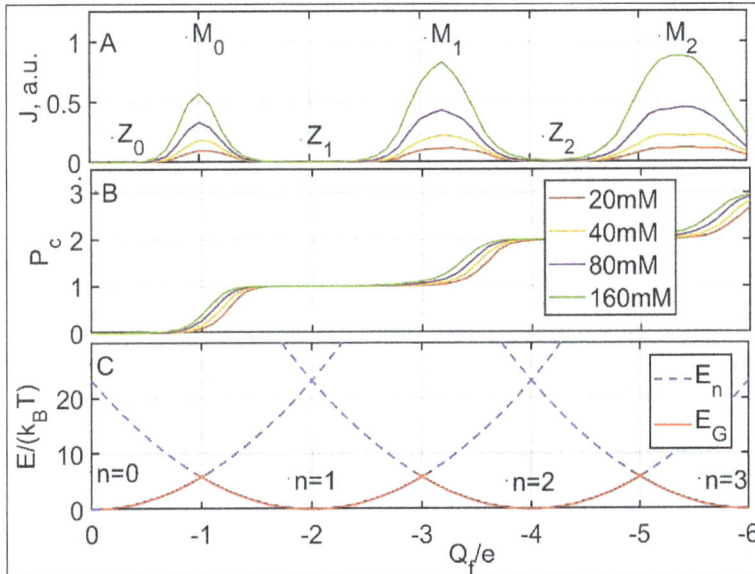

Ionic Coulomb blockade illustrated by BD-simulations of Ca²⁺ conduction, as the fixed charge Q_f is varied: (a) Ca²⁺ conduction bands; (b) Ca²⁺ occupancy, forming a Coulomb staircase; and (c) Ground state energy (red).

The ICB model explicitly predicts an oscillatory dependence of conduction on Q_f, with two interlaced sets of singularities associated with a sequentially increasing number of ions $n = 1, 2, 3, ...$ in the channel.

Electrostatic blockade points Z_n correspond to minima in the ground state energy of the pore.

$$E_G(Q_f) = \min_n E_n(Q_f) \qquad \text{(Ground state)}$$

The Z_n points ($\partial E_n / \partial Q_f = 0$) are equivalent to neutralisation points where $Q_n = 0$.

Resonant conduction points M_n. correspond to the barrier-less condition: $\mu_{ex} = 0$, or $\Delta E \approx -E_{AFF}$.

The values of Z_n and M_n are given by the simple formulae:

$$Z_n = -zen \qquad \text{(Electrostatic blockade)}$$
$$M_n = -ze(n + 1/2) \quad \text{(Resonant conduction)},$$

i.e. the period of conductance oscillations in Q_f, $\Delta = |M_{n+1} - M_n| = |Z_{n+1} - Z_n| = |ze|$.

For $z = 2$, in a typical ion channel geometry, $\Delta E / (k_B T) \approx 20 \gg 1$, and ICB becomes strong. Consequently, plots of the BD-simulated Ca^2+ current J vs Q_f exhibit multi-ion conduction bands - *strong Coulomb blockade oscillations* between minima Z_n and maxima M_n.

The point $Z_0 = 0$ corresponds to an uncharged pore with $Q_f = 0$. Such pores are blockaded for ions of either sign.

Coulomb Staircase

The ICB oscillations in conductance correspond to a *Coulomb staircase* in the pore occupancy P_c, with transition regions corresponding to M_n and saturation regions corresponding to Z_n. The shape of the staircase is described by the Fermi-Dirac (FD) distribution, similarly to the Coulomb staircases of quantum dots. Thus, for the $0 \rightarrow 1$ transition, the FD function is:

$$P_c = \left[1 + \frac{1}{P_b} \exp\left(\frac{\mu_{ex}}{k_B T}\right)\right]^{-1} ; \qquad \text{(Fermi-Dirac distribution)}$$

$$\mu_{ex} = \frac{ze}{C_s}(Q_f - M_0).$$

Here μ_{ex} is the excess chemical potential for the particular ion and P_b is an equivalent bulk occupancy related to pore volume. The saturated FD statistics of occupancy is equivalent to the Langmuir isotherm or to Michaelis-Menten kinetics.

It is the factor $1/P_b$ that gives rise to the concentration-related shift in the staircase.

Shift of Singular Points

Addition of the partial excess chemical potentials μ_{ex}^Y coming from different sources Y (including dehydration, local binding, volume exclusion etc.) leads to the ICB barrier-less condition $\mu_{\text{ex}} = 0$ leads to a proper shift in the ICB resonant points M_n, described by a "shift equation":

$$\Delta M_n = -\frac{C_s}{ze}\sum_Y \mu_{\text{ex}}^Y \quad \text{(Shift equation)}$$

i.e. the additional energy contributions μ_{ex}^Y lead to shifts in the resonant barrier-less point M_0.

The more important of these shifts (excess potentials) are:

- A concentration-related shift $\mu_{\text{ex}}^{\text{ES}} = -k_B T \log(P_b)$ arising from the bulk entropy.

- A dehydration-related shift $\mu_{\text{ex}}^{\text{DH}}$, arising from partial dehydration penalty.

- A local binding-related shift $\mu_{\text{ex}}^{\text{INT}}$, coming from energy of local binding and surface effects.

In Artificial Nanopores

Sub-nm MoS₂ Pores

Following its prediction based on analytic theory and molecular dynamics simulations, experimental evidence for ICB emerged from experiments on monolayer MoS_2 pierced by a single $0.6\,\text{nm}$ nanopore. Highly non-Ohmic conduction was observed between aqueous ionic solutions on either side of the membrane. In particular, for low voltages across the membrane, the current remained close to zero, but it rose abruptly when a threshold of about $400\,\text{mV}$ was exceeded. This was interpreted as complete ionic Coulomb blockade of current in the (uncharged) nanopore due to the large potential barrier at low voltages. But the application of larger voltages pulled the barrier down, producing accessible states into which transitions could occur, thus leading to conduction.

In Biological Ion Channels

The realisation that ICB could occur in biological ion channels accounted for several experimentally observed features of selectivity, including:

Valence Selectivity

Valence selectivity is the channel's ability to discriminate between ions of different valence z, wherein e.g. a calcium channel favours Ca^{2+} ions over Na^+ ions by a factor of up to 1000×. Valence selectivity has been attributed variously to pure electrostatics, or to a charge space competition mechanism, or to a snug fit of the ion to ligands, or to quantised dehydration. In the ICB model, valence selectivity arises from electrostatics, namely from z-dependence of the value of $Q_f = M_n = -ze(n+1/2)$ needed to provide for barrier-less conduction.

Correspondingly, the ICB model provides explanations of why site-directed mutations that alter Q_f can destroy the channel by blockading it, or can alter its selectivity from favouring Ca^{2+} ions to favouring Na^+ ions, or *vice versa*.

Divalent Blockade

Divalent (e.g. Ca^{2+}) blockade of monovalent (e.g. Na^+) currents is observed in some types of ion channels. Namely, Na^+ ions in a pure sodium solution pass unimpeded through a calcium channel, but are blocked by tiny (nM) extracellular concentrations of Ca^{2+} ions. ICB provides a transparent explanation of both the phenomenon itself and of the Langmuir-isotherm-shape of the current *vs.* log[Ca^{2+}] attenuation curve, deriving them from the strong affinity and an FD distribution of Ca^{2+} ions. *Vice versa*, appearance divalent blockade presents strong evidence in favour of ICB.

Similarly, ICB can account for the divalent (Iodide I^{2-}) blockade that has been observed in biological chloride (Cl^-)-selective channels.

Special Features

Comparisons between ICB and ECB

ICB and ECB should be considered as two versions of the same fundamental electrostatic phenomenon. Both ICB and ECB are based on charge quantisation and on the finite single-particle charging energy ΔE, resulting in close similarity of the governing equations and manifestations of these closely related phenomena. Nonetheless, there are important distinctions between ICB and ECB: their similarities and differences are summarised in table.

Comparison between ICB and ECB		
Property	ICB	ECB
Mobile charge carriers	cations (Na^+, K^+, Ca^{2+} etc...), anions (Cl^-, I^{2-} etc.)	electrons (e^-)
Valence of mobile charge carriers	positive (+1, +2, +3,...), negative (−1, −2...)	$z = -1$
Transport engine	Classical diffusion	QM tunneling
Conductance oscillations	Yes, valence dependent	Yes
Coulomb staircase for occupancy, P_c	Yes, FD-shaped	Yes, FD-shaped

Particular Cases

Coulomb blockade can also appear in superconductors; in such a case the free charge carriers are Cooper pairs ($z = -2$).

In addition, Pauli spin blockade represents a special kind of Coulomb blockade, connected with Pauli exclusion principle.

Quantum Analogies

Despite appearing in completely classical systems, ICB exhibits some phenomena reminiscent of quantum-mechanics (QM). They arise because the charge/entity discreteness of the ions leads to quantisation of the energy ΔE spectrum and hence to the QM-analogies:

- Noise-driven diffusive motion provides for escape over barriers, comparable to QM-tunnelling in ECB.

- The particular FD shape of the Ca^{2+} occupancy *vs* $\log[Ca^{2+}]$ plays a significant role in the ICB explanation of the divalent blockade phenomenon. The appearance of an FD distribution in the diffusion of classical particles obeying an exclusion principle, has been demonstrated rigorously.

NANOCIRCUITRY

Nanocircuits are electrical circuits operating on the nanometer scale. This is well into the quantum realm, where quantum mechanical effects become very important. One nanometer is equal to 10^{-9} meters or a row of 10 hydrogen atoms. With such progressively smaller circuits, more can be fitted on a computer chip. This allows faster and more complex functions using less power. Nanocircuits are composed of three different fundamental components. These are transistors, interconnections, and architecture, all fabricated on the nanometer scale.

Various Approaches to Nanocircuitry

A variety of proposals have been made to implement nanocircuitry in different forms. These include Nanowires, Single-Electron Transistors, Quantum dot cellular automata, and Nanoscale Crossbar Latches. However, likely nearer-term approaches will involve incorporation of nanomaterials to improve MOSFETs. These currently form the basis of most analog and digital circuit designs, the scaling of which drives Moore's Law. A review article covering the MOSFET design and its future was published in 2004 comparing different geometries of MOSFETs under scale reduction and noted that circular cross-section vertical channel FETs are optimal for scale reduction. This configuration is capable of being implemented with a high density using vertical semiconductor cylindrical channels with nanoscale diameters and Infineon Technologies and Samsung have begun research and development in this direction resulting in some basic patents using nanowires and carbon nanotubes in MOSFET designs. In an alternative approach, Nanosys uses solution based deposition and alignment processes to pattern pre-fabricated arrays of nanowires on a substrate to serve as a lateral channel of an FET. While not capable of the same scalability as single nanowire FETs, the use of pre-fabricated multiple nanowires for the channel increases reliability and reduces production costs since large volume printing processes may be used to deposit the nanowires at a lower temperature than conventional fabrication procedures. In addition, due to the lower temperature deposition a wider variety of materials such as polymers may be used as the carrier substrate for the transistors opening the door to flexible electronic applications such as electronic paper, bendable flat panel displays, and wide area solar cells.

Production Methods

One of the most fundamental concepts to understanding nanocircuits is the formulation of

Moore's Law. This concept arose when Intel co-founder Gordon Moore became interested in the cost of transistors and trying to fit more onto one chip. It relates that the number of transistors that can be fabricated on a silicon integrated circuit—and therefore the computing abilities of such a circuit—is doubling every 18 to 24 months. The more transistors one can fit on a circuit, the more computational abilities the computer will have. This is why scientists and engineers are working together to produce these nanocircuits so increasingly more and more transistors will be able to fit onto a chip. Despite how good this may sound, there are many problems that arise when so many transistors are packed together. With circuits being so tiny, they tend to have more problems than larger circuits, more particularly heat - the amount of power applied over a smaller surface area makes heat dissipation difficult, this excess heat will cause errors and can destroy the chip. Nanoscale circuits are more sensitive to temperature changes, cosmic rays and electromagnetic interference than today's circuits. As more transistors are packed onto a chip, phenomena such as stray signals on the chip, the need to dissipate the heat from so many closely packed devices, tunneling across insulation barriers due to the small scale, and fabrication difficulties will halt or severely slow progress. Many believe the market for nanocircuits will reach equilibrium around 2015. At this time they believe the cost of a fabrication facility may be as much as $200 billion. There will be a time when the cost of making circuits even smaller will be too much, and the speed of computers will reach a maximum. For this reason, many scientists believe that Moore's Law will not hold forever and will soon reach a peak, since Moore's law is largely predicated on computational gains caused by improvements in micro-lithographic etching technologies.

In producing these nanocircuits, there are many aspects involved. The first part of their organization begins with transistors. As of right now, most electronics are using silicon-based transistors. Transistors are an integral part of circuits as they control the flow of electricity and transform weak electrical signals to strong ones. They also control electric current as they can turn it on off, or even amplify signals. Circuits now use silicon as a transistor because it can easily be switched between conducting and nonconducting states. However, in nanoelectronics, transistors might be organic molecules or nanoscale inorganic structures. Semiconductors, which are part of transistors, are also being made of organic molecules in the nano state.

The second aspect of nanocircuit organization is interconnection. This involves logical and mathematical operations and the wires linking the transistors together that make this possible. In nanocircuits, nanotubes and other wires as narrow as one nanometer are used to link transistors together. Nanowires have been made from carbon nanotubes for a few years. Until a few years ago, transistors and nanowires were put together to produce the circuit. However, scientists have been able to produce a nanowire with transistors in it. In 2004, Harvard University nanotech pioneer Charles Lieber and his team have made a nanowire—10,000 times thinner than a sheet of paper—that contains a string of transistors. Essentially, transistors and nanowires are already pre-wired so as to eliminate the difficult task of trying to connect transistors together with nanowires.

The last part of nanocircuit organization is architecture. This has been explained as the overall way the transistors are interconnected, so that the circuit can plug into a computer or other system and operate independently of the lower-level details. With nanocircuits being so small, they are destined for error and defects. Scientists have devised a way to get around this. Their architecture combines circuits that have redundant logic gates and interconnections with the ability to reconfigure structures at several levels on a chip. The redundancy lets the circuit identify problems and reconfigure itself so the circuit can avoid more problems. It also allows for errors within the logic gate and still have it work properly without giving a wrong result.

Potential Applications and Breakthroughs

Scientists in India have developed the world's smallest transistor which will be used for nanocircuits. The transistor is made entirely from carbon nanotubes. Nanotubes are rolled up sheets of carbon atoms and are more than a thousand times thinner than human hair. Normally circuits use silicon-based transistors, but these will soon replace those. The transistor has two different branches that meet at a single point, hence giving it a Y shape. Current can flow throughout both branches and is controlled by a third branch that turns the voltage on or off. This new breakthrough can now allow for nanocircuits to hold completely to their name as they can be made entirely from nanotubes. Before this discovery, logic circuits used nanotubes, but needed metal gates to be able to control the flow of electric current.

Arguably the biggest potential application of nanocircuits deals with computers and electronics. Scientists and engineers are always looking to make computers faster. Some think in the nearer term, we could see hybrids of micro- and nano-silicon with a nano core—perhaps a high-density computer memory that retains its contents forever. Unlike conventional circuit design, which proceeds from blueprint to photographic pattern to chip, nanocircuit design will probably begin with the chip—a haphazard jumble of as many as 1024 components and wires, not all of which will even work—and gradually sculpt it into a useful device. Instead of taking the traditional top-down approach, the bottom-up approach will probably soon have to be adopted because of the sheer size of these nanocircuits. Not everything in the circuit will probably work because at the nano level, nanocircuits will be more defective and faulty because of their compactness. Scientists and engineers have created all of the essential components of nanocircuits such as transistors, logic gates and diodes. They have all been constructed from organic molecules, carbon nanotubes and nanowire semiconductors. The only thing left to do is find a way to eliminate the errors that come with such a small device and nanocircuits will become a way of all electronics. However, eventually there will be a limit as to how small nanocircuits can become and computers and electronics will reach their equilibrium speeds.

NANORADIO

A nanoradio (also called carbon nanotube radio) is a nanotechnology acting as a radio transmitter and receiver by using carbon nanotubes. One of the first nanoradios was constructed in 2007 by researchers under Alex Zettl at the University of California, Berkeley where they successfully transmitted an audio signal. Due to the small size, nanoradios can have several possible applications such as radio function in the bloodstream.

Properties

The small size, roughly 10 nanometers wide and hundreds of nanometers long, and composition of nanoradios provide several distinct properties. The small size of nanoradios enables electrons to pass through without much friction, making nanoradios efficient conductors. Nanoradios can also come in different sizes; they can be double-walled, tripled-walled and multi-walled. Aside from the different sizes, nanoradios can also take different shapes such as bent, straight or toroidal. Common among all nanoradios is how relatively strong they are. The resistance can be attributed to the strength of the bonds between carbon atoms.

Function

The fundamental parts of a radio are the antenna, tuner, demodulator and amplifier. Carbon nanotubes are special in that they can function as these parts without the need of extra circuitry.

Antenna

The nanoradio is small enough for electromagnetic signals to mechanically vibrate the nanoradio. The nanoradio essentially acts as an antenna by vibrating with the same frequency as the signal from incoming electromagnetic waves; this is in contrast with traditional radio antennas, which are generally stationary. The nanotube can vibrate in high frequencies, from "thousands to millions of times per second."

Tuner

The nanoradio can also function as a tuner by extending or reducing the length of the nanotube; doing so changes the resonance frequency at which it vibrates, enabling the radio to tune into specific frequencies. The length of the nanotube can be extended by pulling the tip with a positive electrode and can be shortened by removing atoms off the tip. Consequently, changing the length is permanent and can't be reversed; however, the method of varying the electric field can also affect the frequency that the nanoradio responds without being permanent.

Amplifier

As a benefit of the microscopic size and needle-like shape, the nanoradio functions naturally as an amplifier. The nanoradio exhibits field emission, in which a small voltage emits a flow of electrons; due to this, a small electromagnetic wave would produce a large flow of electrons, amplifying the signal.

Demodulator

Demodulation is essentially the separation of the information signal from the carrier wave. When the nanoradio vibrates in sync with the carrier wave, the nanoradio responds only to the information signal and ignores the carrier wave; and so, the nanoradio can act as a demodulator without the need of circuitry.

Medical Application

Currently, chemotherapy uses chemicals that harm not only cancerous cells, but also healthy ones since they are put into the blood stream. Nanoradios can be used to prevent damage to healthy cells by remotely communicating with the radio to release drugs and specifically target cancerous cells. Nanoradios can also be injected into individual cells to release certain chemicals, enabling repair of specific cells. Nanoradios can also be used to monitor insulin levels of diabetes patients and use that information to release a drug or chemical.

Complications

The implanting of nanoradios in the body are currently not feasible due to power dissipation. The nanoradio radiates about 4.5×10^{-27} W of electromagnetic power; however, much of this power would be lost when passing through the body. The amount of energy input can be increased, but would generate too much heat in the body, posing a safety risk. Other issues include the difficulty of constructing a nanoradio due to its nanoscale size, requiring quantum models and precision to manufacture.

NANOMOTOR

A nanomotor is a molecular or nanoscale device capable of converting energy into movement. It can typically generate forces on the order of piconewtons.

While nanoparticles have been utilized by artists for centuries, such as in the famous Lycurgus cup, scientific research into nanotechnology did not come about until recently. In 1959, Richard Feynman gave a famous talk entitled "There's Plenty of Room at the Bottom" at the American Physical Society's conference hosted at Caltech. He went on to wage a scientific bet that no one person could design a motor smaller than 400 μm on any side. The purpose of the bet (as with most scientific bets) was to inspire scientists to develop new technologies, and anyone who could develop a nanomotor could claim the $1,000 USD prize. However, his purpose was thwarted by William McLellan, who fabricated a nanomotor without developing new methods. Nonetheless, Richard Feynman's speech inspired a new generation of scientists to pursue research into nanotechnology.

Nanomotors are the focus of research for their ability to overcome microfluidic dynamics present at low Reynold's numbers. Scallop Theory explains that nanomotors must break symmetry to produce motion at low Reynold's numbers. In addition, Brownian motion must be considered because particle-solvent interaction can dramatically impact the ability of a nanomotor to traverse through a liquid. This can pose a significant problem when designing new nanomotors. Current nanomotor research seeks to overcome these problems, and by doing so, can improve current microfluidic devices or give rise to new technologies.

Nanotube and Nanowire Motors

In 2004, Ayusman Sen and Thomas E. Mallouk fabricated the first synthetic and autonomous nanomotor. The two-micron long nanomotors were composed of two segments, platinum and gold, that could catalytically react with diluted hydrogen peroxide in water to produce motion. The Au-Pt nanomotors have autonomous, non-Brownian motion that stems from the propulsion via catalytic generation of chemical gradients. As implied, their motion does not require the presence of an external magnetic, electric or optical field to guide their motion. By creating their own local fields, these motors are said to move through self-electrophoresis. Joseph Wang in 2008 was able to dramatically enhance the motion of Au-Pt catalytic nanomotors by incorporating carbon nanotubes into the platinum segment.

Since 2004, different types of nanotube and nanowire based motors have been developed, in addition to nano- and micromotors of different shapes. Most of these motors use hydrogen peroxide as fuel, but some notable exceptions exist.

These silver halide and silver-platinum nanomotors are powered by halide fuels, which can be regenerated by exposure to ambient light. Some nanomotors can even be propelled by multiple stimuli, with varying responses. These multi-functional nanowires move in different directions depending on the stimulus (e.g. chemical fuel or ultrasonic power) applied. For example, bimetallic nanomotors have been shown to undergo rheotaxis to move with or against fluid flow by a combination of chemical and acoustic stimuli. In Dresden Germany, rolled-up microtube nanomotors produced motion by harnessing the bubbles in catalytic reactions. Without the reliance on electrostatic interactions, bubble-induced propulsion enables motor movement in relevant biological fluids, but typically still requires toxic fuels such as hydrogen peroxide. This has limited nanomotors' in vitro applications. One in vivo application, however, of microtube motors has been described for the first time by Joseph Wang and Liangfang Zhang using gastric acid as fuel. Future research into catalytical nanomotors holds major promise for important cargo-towing applications, ranging from cell sorting microchip devices to directed drug delivery.

Enzymatic Nanomotors

Recently, there has been more research into developing enzymatic nanomotors and micropumps. At low Reynold's numbers, single molecule enzymes could act as autonomous nanomotors. Ayusman Sen and Samudra Sengupta demonstrated how self-powered micropumps can enhance particle transportation. This proof-of-concept system demonstrates that enzymes can be successfully utilized as an "engine" in nanomotors and micropumps. It has since been shown that particles themselves will diffuse faster when coated with active enzyme molecules in a solution of their substrate. Further, it has been seen through microfluidic experiments that enzyme molecules will undergo directional swimming up their substrate gradient. This remains the only method of separating enzymes based on activity alone. Additionally, enzymes in cascade have also shown aggregation based on substrate driven chemotaxis. Developing enzyme-driven nanomotors promises to inspire new biocompatible technologies and medical applications.

A proposed branch of research is the integration of molecular motor proteins found in living cells into molecular motors implanted in artificial devices. Such a motor protein would be able to move a "cargo" within that device, via protein dynamics, similarly to how kinesin moves various molecules along tracks of microtubules inside cells. Starting and stopping the movement of such motor proteins would involve caging the ATP in molecular structures sensitive to UV light. Pulses of UV illumination would thus provide pulses of movement. DNA nanomachines, based on changes between two molecular conformations of DNA in response to various external triggers.

Helical Nanomotors

Another interesting direction of research has led to the creation of helical silica particles coated with magnetic materials that can be maneuvered using a rotating magnetic field.

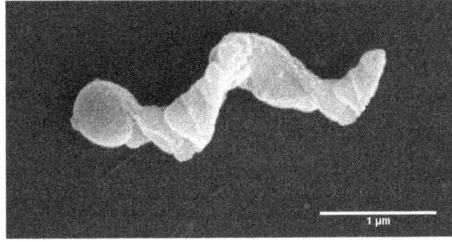

Scanning Electron Microscope image of a Helical nanomotor.

Such nanomotors are not dependent on chemical reactions to fuel the propulsion. A triaxial Helmholtz coil can provide directed rotating field in space. Recent works have shown how such nanomotors can be used to measure viscosity of non-newtonian fluids at a resolution of a few microns. This technology promises creation of viscosity map inside cells and the extracellular milieu. Such nanomotors have been demonstrated to move in blood. Recently, researchers have managed to controllably move such nanomotors inside cancer cells allowing them to trace out patterns inside a cell.

NANOIONIC DEVICE

Nanoionic hard drives use nanoionic technology allowing for smaller devices while doing away with moving parts and the mechanical failures which are associated with previous HDD drives. Nanoionic hard drives are currently the most state of the art drives on the market and nanoionics was not utilized in hard drives until February 2014. Nanoionic devices were first proposed in 1992: "The results obtained show that it is possible to form arrays of electrochemical devices with single elements ~10 nm in size in the films". The basis of design of nanoionic devices is the creation of nanostructures with nanoionic parameter $\lambda / L \sim 1$, where L is the size of device structure, and λ is the characteristic size of specific region where the property of fast ionic transport is realized. "Possibilities to influence on these specific regions $< \lambda >$ in a controllable manner may appear in short sized devices". Ion - electronic hybrid devices should be considered as a step on a way to the future nanoelectronics-nanoionics (nanoelionics) that was first proposed in 1996.

Comparison

Traditional hard drives run on mechanical parts, which uses the permanent disk to store your personal data. The hard disk drives usually have a 10 centimeter diameter and are a centimeter thick. These drives use a magnetic recording technique. We, as users, can easily manipulate our data on this medium i.e. add, copy, cut or paste data on it though this technique. The principle of magnetic flux is used for such processes. Basically the hard drive remembers which data is where and how to manipulate the data according to the user's preference by remembering the flux patterns. The average hard drives contain the specific parts:

- Disk case - This is a rectangular shaped disk case (the main part of the drive) which contains all the components screwed together.

- Disk platter – This is the actual disk where the user's data is recorded on. Consists of patterns on which the computer navigates through the data using the principle of magnetic flux.

- Spindle – This component holds the magnetic platters together and is responsible for spinning the disk platter when required.

- Read/Write Arm – This component acts as the navigator. It navigates through the magnetic patterns looking for the particular data.

SSD, regarded as the future of data storage, is a technology recently available to the consumers. However expensive, these state of the art drives are considerably smaller. Data on such drives is statically twice as safer i.e. such drives crash less. First of all, these drives do not involve magnetic principles. These drives use semiconductors to store data unlike the disk platter used in the Hard Disk Drives. SSDs use the principle of flash drive. It consists of no actual mechanical parts i.e. no movements of parts is involved.

Mode of Operation

Nanoionics devices rely on the fundamentals of electrochemistry. Currently the hard drives are made up of solid materials which weigh less and generate more power; solids have one polarity for ions that are moving. The ions are what the electrodes are made of. Electrodes are key components in nanoionic devices. Those electrodes can be made out of ZrO_2, a metal which is coated in La_2NiO_4/La_2CuO_4, or $Bi_{10}V_4(metal)O_{26}$. Where "metal" is any metal found in the transition metal group like copper. These nanoionic devices are made up of smaller devices that are that are spaced less than a tenth nanometer away. Due to the small spaced distance between the materials in the nanoionic device smaller ions are necessary for the pathway.Metals in the first group of the periodic table are small, but they are too reactive. So there needs to be a compromise between the reactivity of chemicals and size. That is why materials like Cu or Ag fulfill the needs of the nanoionic drives.

In the nanoionic device there would be chalcogenide glass with has the metal like gold and an element for group six infused in the glass. This glass is the electrode for the nanoionic. In the oxidation reaction the Ag^+ then loses the electron and turns into Ag. This reaction only happens when a switch confirms the reaction to happen and these switches are used for the binary information storage. That binary information storage is where all the data is saved on the hard drive. All this is dependent on the small current of ions, hence the name nanoionic, to allow the reaction to happen. All the components of the switch, right metals, and solid have to come together to make the nanoionic happen correctly.

Problems

Due to nanoionic devices being new technology, they lack memory capacity. Smaller drives take up less charge to change the electrolytes in their nonvolatile state. The smaller size is the only way the nanoionic devices are feasible with today's technology because it needs the reduction in resistance. These smaller sizes can hold a lot of memory and can still work in a computer.

NANOBATTERIES

Nanobatteries are fabricated batteries employing technology at the nanoscale, particles that measure less than 100 nanometers or 10^{-7} meters. These batteries may be nano in size or may use nanotechnology in a macro scale battery. Nanoscale batteries can be combined together to function as a macrobattery such as within a nanopore battery.

Traditional lithium-ion battery technology uses active materials, such as cobalt-oxide or manganese oxide, with particles that range in size between 5 and 20 micrometers (5000 and 20000 nanometers – over 100 times nanoscale). It is hoped that nano-engineering will improve many of the shortcomings of present battery technology, such as volume expansion and power density.

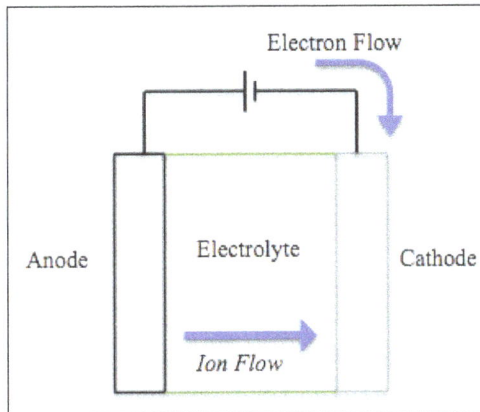

A basic schematic of how an ion battery works. The blue arrows indicate discharging. If both arrows reversed direction, the battery would be charging and this battery would then be considered a secondary (rechargeable) battery.

A battery converts chemical energy to electrical energy and is composed of three general parts:

- Anode (positive electrode).

- Cathode (negative electrode).

- Electrolyte.

The anode and cathode have two different chemical potentials, which depend on the reactions that occur at either terminus. The electrolyte can be a solid or a liquid, referring to a dry cell or wet cell respectively and is ionically conductive. The boundary between the electrode and electrolyte is called the solid-electrolyte interphase (SEI). An applied voltage across the electrodes causes the chemical energy stored in the battery to be converted to electrical energy.

Limitations of Current Battery Technology

A battery's ability to store charge is dependent on its energy density and power density. It is important that charge can remain stored and that a maximum amount of charge can be stored within a battery. Cycling and volume expansion are also important considerations as well. While many other types of batteries exist, current battery technology is based on lithium-ion intercalation technology for its high power and energy densities, long cycle life and no memory effects. These characteristics have led lithium-ion batteries to be preferred over other battery types. To improve a battery technology, cycling ability and energy and power density must be maximized and volume expansion must be minimized.

During lithium intercalation, the volume of the electrode expands, causing mechanical strain.

The mechanical strain compromises the structural integrity of the electrode, causing it to crack. Nanoparticles can decrease the amount of strain placed on a material when the battery undergoes cycling, as the volume expansion associated with nanoparticles is less than the volume expansion associated with microparticles. The little volume expansion associated with nanoparticles also improves the reversibility capability of the battery: the ability of the battery to undergo many cycles without losing charge.

In current lithium-ion battery technology, lithium diffusion rates are slow. Through nanotechnology, faster diffusion rates can be achieved. Nanoparticles require shorter distances for the transport of electrons, which leads to faster diffusion rates and a higher conductivity, which ultimately leads to a greater power density.

Advantages of Nanotechnology

Using nanotechnology to manufacture of batteries offers the following benefits:

- Increasing the available power from a battery and decreasing the time required to recharge a battery. These benefits are achieved by coating the surface of an electrode with nanoparticles, increasing the surface area of the electrode thereby allowing more current to flow between the electrode and the chemicals inside the battery.

- Nanomaterials can be used as a coating to separate the electrodes from any liquids in the battery, when the battery is not in use. In the current battery technology, the liquids and solids interact, causing a low level discharge. This decreases the shelf life of a battery.

Disadvantages of Nanotechnology

Nanotechnology provides its own challenges in batteries:

- Nanoparticles have low density and high surface area. The greater the surface area, the more likely reactions are to occur at the surface with the air. This serves to destabilize the materials in the battery.

- Due to nanoparticles' low density, a higher interparticle resistance exists, decreasing the electrical conductivity of the material.

- Nanomaterials can be difficult to manufacture, increasing their cost. While nanomaterials may greatly improve the abilities of a battery, they may be cost-prohibitive to make.

Active and Past Research

Much research has been performed surrounding lithium ion batteries to maximize their potential. Research into improving batteries is important for clean energy solutions. In order to properly harness clean energy resources, such as solar power, wind power and tidal energy, batteries, capable of storing massive amounts of energy used in grid energy storage, are required. Lithium iron phosphate electrodes are being researched for potential applications to grid energy storage.

Electric vehicles are another technology requiring improved batteries. Electric vehicle batteries currently require large charge times, effectively prohibiting the use for long-distance electric cars.

Nanostructured Anode Materials

Graphite and SEI

The anode in lithium-ion batteries is almost always graphite. Graphite anodes need to improve their thermal stability and create a higher power capability. Graphite and certain other electrolytes can undergo reactions that reduce the electrolyte and create a SEI, effectively reducing the potential of the battery. Nanocoatings at the SEI are currently being researched to stop these reactions from occurring.

In Li-ion batteries, the SEI is necessary for thermal stability, but hinders the flow of lithium ions from the electrode to the electrolyte. Park et al. have developed a nanoscale polydopamine coating such that the SEI no longer interferes with the electrode; instead the SEI interacts with the polydopamine coating.

Graphene and other Carbon Materials

Graphene has been studied extensively for its use in electrochemical systems such as batteries since its first isolation in 2004. Graphene offers high surface area and good conductivity. In current lithium-ion battery technology, the 2D networks of graphite inhibit smooth lithium ion intercalation; the lithium ions must travel around the 2D graphite sheets to reach the electrolyte. This slows the charging rates of the battery. Porous graphene materials are currently being studied to improve this problem. Porous graphene involves either formation of defects in the 2D sheet or the creation of a 3D graphene-based porous superstructure.

As an anode, graphene would provide space for expansion such that the problem of volume expansion does not occur. 3D graphene has shown extremely high lithium ion extraction rates, indicating a high reversible capacity. As well, the random "house-of-cards" visualization seen below of the graphene anode would allow lithium ions to be stored not only on the internal surface of graphene, but also on the nanopores that exist between the single layers of graphene.

Raccichini et al. also outlined the drawbacks of graphene and graphene-based composites. Graphene has a large irreversible mechanism during the first lithiation step. As graphene has a large surface area, this will result in a large initial irreversibility capacity. He proposed that this drawback was so large that graphene-based cells are "unfeasible". Research is still being done on graphene in anodes.

Carbon nanotubes have been used as electrodes for batteries that use intercalation, like lithium-ion batteries, in an effort to improve capacity.

Titanium Oxides

Titanium oxides are another anode material that have been researched for their applications to electric vehicles and grid energy storage. However, low electronic and ionic capabilities, as well as the high cost of titanium oxides have proven this material to be unfavourable to other anode materials.

Silicon-based Anodes

Silicon-based anodes have also been research namely for their higher theoretical capacity than that of graphite. Silicon-based anodes have high reaction rates with the electrolyte, low volumetric capacity and an extremely large volume expansion during cycling. However, recent work has been done to decrease volume expansion in silicon-based anodes. By creating a sphere of conductive carbon around the silicon atom, Liu et al. has proven that this small structural change leaves enough room for the silicon to expand and contract without providing mechanical strain on the electrode.

Nanostructured Cathode Materials

Carbon nanostructures have been used to increase the capability of electrodes, namely the cathode. In $LiSO_2$ batteries, carbon nanostructuring was able to theoretically increase the energy density of the battery by 70% from the current lithium-ion battery technology. In general, lithium alloyshave been found to have an increased theoretical energy density than lithium ions.

Traditionally, $LiCoO_2$ has been used as the cathode in lithium-ion batteries. The first successful alternative cathode for use in electric vehicles has been $LiFeO_4$. $LiFeO_4$ has shown increased power density, a longer lifetime and improved safety over $LiCoO_2$.

Graphene

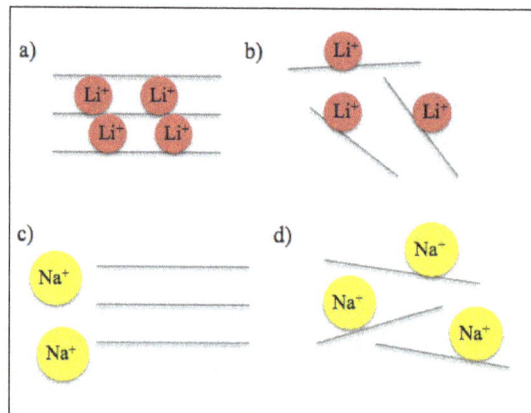

During intercalation, a) lithium ions into a graphite lattice, b) lithium ions into a graphene lattice, c) sodium ions unable to fit into a graphite lattice, d) sodium ions into a graphene lattice.

Graphene could be used to improve the electrical conductivity of cathode materials. $LiCoO_2$, $LiMn_2O_4$ and $LiFePO_4$ are all commonly used cathode materials in lithium-ion batteries. These cathode materials have typically mixed with other carbon-composite materials to improve their rate capability. As graphene has a higher electrical conductivity than these other carbon-composite materials, like carbon black, graphene has a greater ability to improve these cathode materials more than other carbon-composite additives.

Piao et al. has specifically studied porous graphene in relation to just graphene. Porous graphene combined with $LiFePO_4$ was advantageous over just graphene combined with $LiFePO_4$, for improved cycle stability. Porous graphene created good pore channels for the diffusion of lithium ions and prevented the buildup of $LiFePO_4$ particles.

Raccichini et al. suggested graphene-based composites as cathodes in sodium-ion batteries. Sodium ions are too large to fit into the typical graphite lattice, so graphene would allow sodium ions to intercalate. Graphene has also been suggested to fix some of the problems related to lithium-sulphur batteries. Problems associated with lithium sulphur batteries include dissolution of the intermediate in the electrolyte, large volume expansion and poor electrical conductivity. Graphene has been mixed with sulphur at the cathode in an attempt to improve the capacity, stability and conductivity of these batteries.

Conversion Electrodes

Conversion electrodes are electrodes where chemical ionic bonds are broken and reformed. A transformation of the crystalline structure of the molecules also occurs. In conversion electrodes, three lithium ions can be accommodated for every metal ion, whereas the current intercalation technology can only accommodate one lithium ion for every metal ion. Larger lithium to metal ion ratios indicate increased battery capacity. A disadvantage of conversion electrodes is its large voltage hysteresis.

Mapping

Balke et al. is aiming to understand the intercalation mechanism for lithium-ion batteries at the nanoscale. This mechanism is understood at the microscale, but behavior of matter changes depending on the size of the material. Zhu et al. are also mapping the intercalation of lithium ions at the nanoscale using scanning probe microscopy.

Mathematical models for lithium battery intercalation have been calculated and are still under investigation. Whittingham suggested that there was no single mechanism by which lithium ions move through the electrolyte of the battery. The movement depended on a variety of factors including, but not limited to, particle size, the thermodynamic state or metastable state of the battery and whether the reaction operated continuously. Their experimental data for $LiFePO_4 - FePO_4$ suggested the movement of Li-ions in a curved path rather than a linear straight jump within the electrolyte.

Intercalation mechanisms have been studied for polyvalent cations as well. Lee et al. has studied and determined the proper intercalation mechanism for rechargeable zinc batteries.

Stretchable Electronics

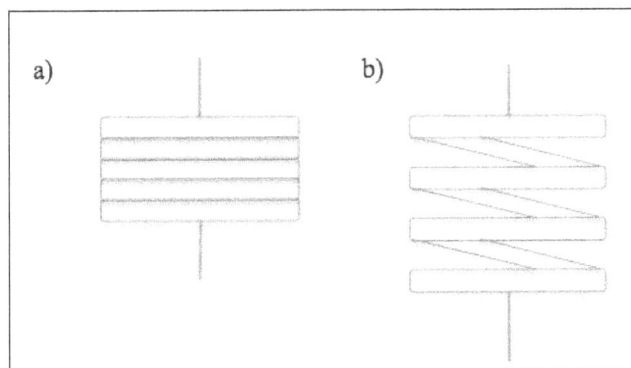

These fiber-like electrodes are wound like springs to get their flexibility. a) is an unstretched spring and b) is a partially stretched spring, showing how pliant these fibers are.

Research has also been done to use carbon nanotube fiber springs as electrodes. $LiMn_2O_4$ and $Li_4Ti_5O_{12}$ are the nanoparticles that have been used as the cathode and anode respectively, and have demonstrated the ability to stretch 300% of their original length. Applications for stretchable electronics include energy storage devices and solar cells.

Printable Batteries

Researchers at the University of California, Los Angeles have successfully developed a "nanotube ink" for manufacturing flexible batteries using printed electronics techniques. A network of carbon nanotubes has been used as a form of electronically conducting nanowires in the cathode of a zinc-carbon battery. Using nanotube ink, the carbon cathode tube and manganese oxide electrolyte components of the zinc-carbon battery can be printed as different layers on a surface, over which an anode layer of zinc foil can be printed. This technology replaces charge collectors like metal sheets or films with a random array of carbon nanotubes. The carbon nanotubes add conductance. Thin and flexible batteries can be manufactured that are less than a millimeter thick.

Although discharge currents of the batteries are at present below the level of practical use, the nanotubes in the ink allow the charge to conduct more efficiently than in a conventional battery, such that the nanotube technology could lead to improvements in battery performance. Technology like this is applicable to solar cells, supercapacitors, light-emitting diodes and smart radio frequency identification (RFID) tags.

Toshiba

By using nanomaterial, Toshiba has increased the surface area of the lithium and widened the bottleneck, allowing the particles to pass through the liquid and recharge the battery more quickly. Toshiba states that it tested a new battery by discharging and fully recharging one thousand times at 77 degrees and found that it lost only one percent of its capacity, an indication of a long battery life.

Toshiba's battery is 3.8 mm thick, 62 mm high and 35 mm deep.

A123Systems

A123Systems has also developed a commercial nano Li-Ion battery. A123 Systems claims their battery has the widest temperature range at -30 to +70 °C. Much like Toshiba's nanobattery, A123 Li-Ion batteries charge to "high capacity" in five minutes. Safety is a key feature touted by the A123 technology, with a video on their website of a nail drive test, in which a nail is driven through a traditional Li-Ion battery and an A123 Li-Ion battery, where the traditional battery flames up and bubbles at one end, the A123 battery simply emits a wisp of smoke at the penetration site. Thermal conductivity is another selling point for the A123 battery, with the claim that the A123 battery offers 4 times higher thermal conductivity than conventional Lithium-Ion cylindrical cells. The nanotechnology they employ is a patented nanophosphate technology.

Valence

Also in the market is Valence Technology, Inc. The technology they are marketing is Saphion Li-Ion Technology. Like A123, they are using a nanophosphate technology, and different active materials than traditional Li-Ion batteries.

Altair

AltairNano has also developed a nanobattery with a one-minute recharge. The advance that Altair claims to have made is in the optimization of nano-structured lithium titanate spinel oxide (LTO).

U.S. Photonics

U.S. Photonics is in the process of developing a nanobattery utilizing "environmentally friendly" nanomaterials for both the anode and cathode as well as arrays of individual nano-sized cell containers for the solid polymer electrolyte. U.S. Photonics has received a National Science Foundation SBIR phase I grant for development of nanobattery technology.

Sony

Produced the first cobalt-based lithium-ion battery in 1991. Since the inception of this first Li-ion battery, the research of nanobatteries has been underway with Sony continuing their strides into the nanobattery field.

NANOGENERATOR

A Nanogenerator is a type of technology that converts mechanical/thermal energy as produced by small-scale physical change into electricity. A Nanogenerator has three typical approaches: piezoelectric, triboelectric, and pyroelectric nanogenerators. Both the piezoelectric and triboelectric nanogenerators can convert mechanical energy into electricity. However, pyroelectric nanogenerators can be used to harvest thermal energy from a time-dependent temperature fluctuation.

Piezoelectric Nanogenerator

A piezoelectric nanogenerator is an energy harvesting device capable of converting external kinetic energy into electrical energy via action by a nano-structured piezoelectric material. Although its definition may include any types of energy harvesting devices using nano-structures to convert various types of ambient energy (e.g. solar power and thermal energy), it is generally used to indicate kinetic energy harvesting devices utilizing nano-scaled piezoelectric material since its first introduction in 2006.

Although still in the early stages of development, the technology has been regarded as a potential breakthrough toward further miniaturization of conventional energy harvesters, possibly leading to facile integration with other types of energy harvesters and the independent operation of mobile electronic devices with reduced concern for sources of energy.

Mechanism

The working principle of nanogenerator will be explained for 2 different cases: the force exerted perpendicular and parallel to the axis of the nanowire.

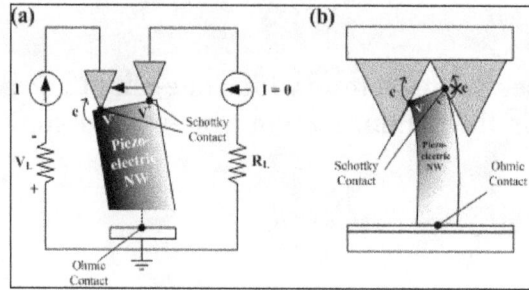

In figure, working principle of nanogenerator where an individual nanowire is subjected to the force exerted perpendicular to the growing direction of nanowire. (a) An AFT tip is swept through the tip of the nanowire. Only negatively charged portion will allow the current to flow through the interface. (b) The nanowire is integrated with the counter electrode with AFT tip-like grating. As of (a), the electrons are transported from the compressed portion of nanowire to the counter electrode because of Schottky contact.

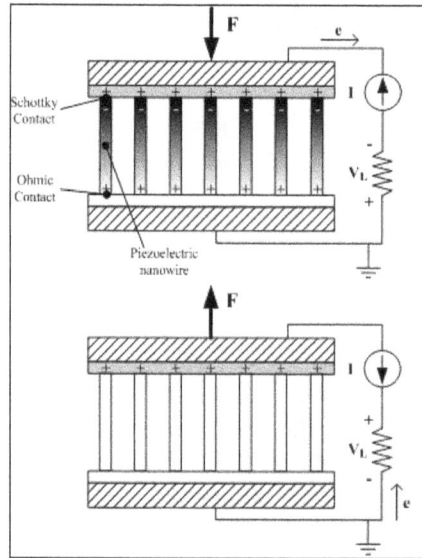

Working principle of nanogenerator where an individual nanowire is subjected to the force exerted parallel to the growing direction of nanowire.

The working principle for the first case is explained by a vertically grown nanowire subjected to the laterally moving tip. When a piezoelectric structure is subjected to the external force by the moving tip, the deformation occurs throughout the structure. The piezoelectric effect will create the electrical field inside the nanostructure; the stretched part with the positive strain will exhibit the positive electrical potential, whereas the compressed part with the negative strain will show the negative electrical potential. This is due to the relative displacement of cations with respect to anions in its crystalline structure. As a result, the tip of the nanowire will have an electrical potential distribution on its surface, while the bottom of the nanowire is neutralized since it is grounded. The maximum voltage generated in the nanowire can be calculated by the following equation:

$$V_{max} = \pm \frac{3}{4(\kappa_0 + \kappa)}[e_{33} - 2(1+v)e_{15} - 2ve_{31}]\frac{a^3}{l^3}v_{max}$$

where, κ_0 is the permittivity in vacuum, κ is the dielectric constant, e_{33}, e_{15} and e_{31} are the piezoelectric coefficients, ν is the Poisson ratio, a is the radius of the nanowire, l is the length of the nanowire and v_{max} is the maximum deflection of the nanowire's tip.

The electrical contact plays an important role to pump out charges in the surface of the tip. The schottky contact must be formed between the counter electrode and the tip of the nanowire since the ohmic contact will neutralize the electrical field generated at the tip. In order to form an effective schottky contact, the electron affinity(E_a) must be smaller than the work function(φ) of the metal composing the counter electrode. For the case of ZnO nanowire with the electron affinity of 4.5 eV, Pt (φ=6.1eV) is a suitable metal to construct the schottky contact. By constructing the schottky contact, the electrons will pass to the counter electrode from the surface of the tip when the counter electrode is in contact with the regions of the negative potential, whereas no current will be generated when it is in contact with the regions of the positive potential, in the case of n-type semiconductive nanostructure (p-type semiconductive structure will exhibit the reversed phenomenon since the hole is mobile in this case). The formation of the schottky contact also contributes to the generation of direct current output signal consequently.

For the second case, a model with a vertically grown nanowire stacked between the ohmic contact at its bottom and the schottky contact at its top is considered. When the force is applied toward the tip of the nanowire, the uniaxial compressive is generated in the nanowire. Due to the piezoelectric effect, the tip of the nanowire will have a negative piezoelectric potential, increasing the Fermi level at the tip. Since the electrons will then flow from the tip to the bottom through the external circuit as a result, the positive electrical potential will be generated at the tip. The schottky contact will barricade the electrons being transported through the interface, therefore maintaining the potential at the tip. As the force is removed, the piezoelectric effect diminishes, and the electrons will be flowing back to the top in order to neutralize the positive potential at the tip. The second case will generate alternating current output signal.

Geometrical Configuration

Depending on the configuration of piezoelectric nanostructure, the most of the nanogenerator can be categorized into 3 types: VING, LING and "NEG". Still, there is a configuration that do not fall into the aforementioned categories, as stated in other type.

Vertical Nanowire Integrated Nanogenerator (VING)

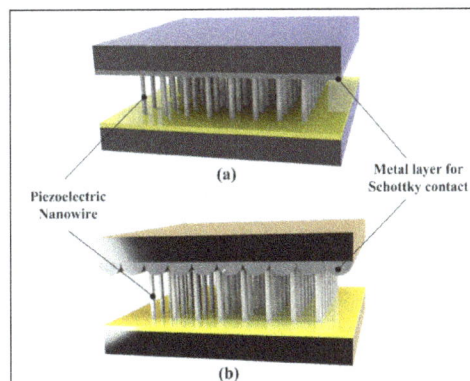

Schematic view of typical Vertical nanowire Integrated Nanogenerator, (a) with full contact, and (b) with partial contact. Note that the grating on the counter electrode is important in the latter case.

Ving is a 3-dimensional configuration consisting of a stack of 3 layers in general, which are the base electrode, the vertically grown piezoelectric nanostructure and the counter electrode. The piezoelectric nanostructure is usually grown from the base electrode by various synthesizing techniques, which are then integrated with the counter electrode in full or partial mechanical contact with its tip.

After Professor Zhong Lin Wang of the Georgia Institute of Technology has introduced a basic configuration of VING in 2006 where he used a tip of atomic force microscope (AFM) to induce the deformation of a single vertical ZnO nanowire, the first development of VING is followed in 2007. The first VING utilizes the counter electrode with the periodic surface grating resembling the arrays of AFM tip as a moving electrode. Since the counter electrode is not in full contact with the tips of the piezoelectric nanowire, its motion in-plane or out-of-plane occurred by the external vibration induces the deformation of the piezoelectric nanostructure, leading to the generation of the electrical potential distribution inside each individual nanowire. It should be noted that the counter electrode is coated with the metal forming the schottky contact with the tip of the nanowire, where only the compressed portion of piezoelectric nanowire would allow the accumulated electrons pass through the barrier between its tip and the counter electrode, in case of n-type nanowire. The switch-on and –off characteristic of this configuration shows its capability of generating direct current generation without any requirement for the external rectifier.

In VING with partial contact, the geometry of the counter electrode plays an important role. The flat counter electrode would not induce the sufficient deformation of the piezoelectric nanostructures, especially when the counter electrode moves by in-plane mode. After the basic geometry resembling the array of AFM tips, a few other approaches have been followed for facile development of the counter electrode. Professor Zhong Lin Wang's group have generated counter electrode composed of ZnO nanorods utilizing the similar technique used for synthesizing ZnO nanowire array. Professor Sang-Woo Kim's group of Sungkyunkwan University (SKKU) and Dr. Jae-Young Choi's group of Samsung Advanced Institute of Technology (SAIT) in South Korea introduced bowl-shaped transparent counter electrode by combining anodized aluminum and the electroplating technology. They also have developed the other type of the counter electrode by using networked single-walled carbon-nanotube (SWNT) on the flexible substrate, which is not only effective for energy conversion but also transparent.

The other type of VING has been also suggested. While it shares the identical geometric configuration with the aforementioned, such a VING has full mechanical contact between the tips of the nanowires and the counter electrode. This configuration is effective for application where the force is exerted in the vertical direction (toward the c axis of the piezoelectric nanowire), and it generates alternating current (AC) unlike VINGs with partial contact.

Lateral Nanowire Integrated Nanogenerator (LING)

Ling is a 2-dimensional configuration consisting of three parts: the base electrode, the laterally grown piezoelectric nanostructure and the metal electrode for schottky contact. In most of cases, the thickness of the substrate film is much thicker than the diameter of the piezoelectric nanostructure, so the individual nanostructure is subjected to the pure tensile strain.

LING is an expansion of single wire generator (SWG), where a laterally aligned nanowire is integrated on the flexible substrate. SWG is rather a scientific configuration used for verifying the capability of electrical energy generation of a piezoelectric material and is widely adopted in the early stage of the development.

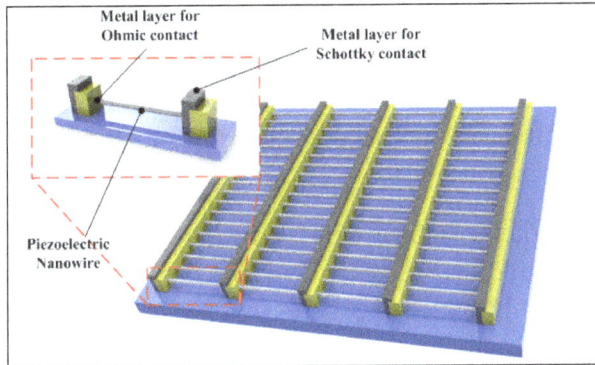

Schematic view of typical Lateral nanowire Integrated Nanogenerator.

As of VINGs with full mechanical contact, LING generates AC electrical signal. The output voltage can be amplified by constructing an array of LING connected in series on the single substrate, leading the constructive addition of the output voltage. Such a configuration may lead to the practical application of LING for scavenging large-scale power, for example, wind or ocean waves.

Nanocomposite Electrical Generators (NEG)

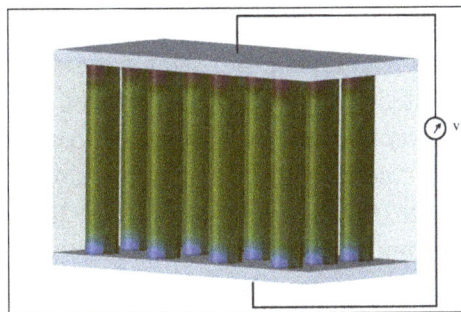

Schematic view of typical Nanocomposite Electrical Generator.

"NEG" is a 3-dimensional configuration consisting three main parts: the metal plate electrodes, the vertically grown piezoelectric nanostructure and the polymer matrix which fills in between in the piezoelectric nanostructure.

NEG was introduced by Momeni et al. It was shown that NEG has a higher efficiency compared to original nanogenerator configuration which a ZnO nanowire will be bended by an AFM tip. It is also shown that it provides an energy source with higher sustainability.

Other type: The fabric-like geometrical configuration has been suggested by Professor Zhong Lin Wang in 2008. The piezoelectric nanowire is grown vertically on the two microfibers in its radial direction, and they are twined to form a nanogenerator. One of the microfibers is coated with the metal to form a schottky contact, serving as the counter electrode of VINGs. As the movable microfiber is stretched, the deformation of the nanostructure occurs on the stationary microfiber, resulting in the voltage generation. Its working principle is identical to VINGs with partial mechanical contact, thus generating DC electrical signal.

Materials

Among various piezoelectric materials studied for the nanogenerator, many of the researches have been focused on the materials with wurtzite structure such as ZnO, CdS and GaN. The greatest advantage of these material arises from the facile and cost-effective fabrication technique, hydrothermal synthesis. Since the hydrothermal synthesis can be conducted in a low temperature environment under 100 °C in addition to vertical and crystalline growth, these materials can be integrated in various substrates with reduced concern for its physical characteristics such as a melting temperature.

Endeavors for enhancing the piezoelectricity of the individual nanowire also led to the development of other piezoelectric materials based on Wurtzite structure. Professor Zhong Lin Wang of Georgia Institute of Technology introduced p-type ZnO nanowire. Unlike the n-type semiconductive nanostructure, the mobile particle in p-type is a hole, thus the schottky behavior is reversed from that of n-type case; the electrical signal is generated from the portion of the nanostructure where the holes are accumulated. It is experimentally proved that p-type ZnO nanowire can generate the output signal near 10 times that of n-type ZnO nanowire.

From the idea that the material with perovskite structure is known to have more effective piezoelectric characteristic compared to that with wurtzite structure, Barium titanate ($BaTiO_3$) nanowire has been also studied by Professor Min-Feng Yu of University of Illinois at Urbana Champaign. The output signal is found to be more than 16 time that from a similar ZnO nanowire.

Professor Liwei Lin of University of California, Berkeley has suggested that PVDF can be also applied to form a nanogenerator. Being a polymer, PVDF utilizes a near-field electrospinning for its fabrication, which is rather a different technique compared to other materials. The nanofiber can be directly written on the substrate controlling the process, and this technique is expected to be applied for forming self-powered textile based on nanofiber. Researchers from SUTD presented the successful synthesis of ultra-long potassium niobate ($KNbO_3$) nanofibers using a sol-gel assisted far-field electrospinning process and utilized them to develop a high output voltage flexible nanogenerator.

Considering that the piezoelectric constant plays a critical role in the overall performance of a piezoelectric nanogenerator, another research direction to improve device efficiency is to find new material of large piezoelectric response. Lead Magnesium Niobate-Lead Titanate (PMN-PT) is a next-generation piezoelectric material with super high piezoelectric constant when ideal composition and orientation are obtained. In 2012, PMN-PT Nanowires with a very high piezoelectric constant were fabricated by a hydro-thermal approach and then assembled into an energy-harvesting device. The record-high piezoelectric constant was further improved by the fabrication of a single-crystal PMN-PT nanobelt, which was then used as the essential building block for a piezoelectric nanogenerator.

Applications

Nanogenerator is expected to be applied for various applications where the periodic kinetic energy exists, such as wind and ocean waves in a large scale to the muscle movement by the beat of a heart or inhalation of lung in a small scale. The further feasible applications are as follows.

1. Self-powered nano/micro devices: One of the feasible applications of nanogenerator is an independent or a supplementary energy source to nano/micro devices consuming relatively low amount of energy in a condition where the kinetic energy is supplied continuously. One of example has been introduced by Professor Zhong Lin Wang's group in 2010 by the self-powered pH or UV sensor integrated VING with an output voltage of 20~40 mV onto the sensor.

Still, the converted electrical energy is relatively small for operating nano/micro devices; therefore the range of its application is still bounded as a supplementary energy source to the battery. The breakthrough is being sought by combining the nanogenerator with the other types of energy harvesting devices, such as solar cell or biochemical energy harvester. This approach is expected to contribute to the development of the energy source suitable for the application where the independent operation is crucial, such as Smartdust.

2. Smart Wearable Systems: The outfit integrated or made of the textiles with the piezoelectric fiber is one of the feasible applications of the nanogenerator. The kinetic energy from the human body is converted to the electrical energy through the piezoelectric fibers, and it can be possibly applied to supply the portable electronic devices such as health-monitoring system attached with the Smart Wearable Systems. The nanogenerator such as VING can be also easily integrated in the shoe employing the walking motion of human body.

Another similar application is a power-generating artificial skin. Professor Zhong Lin Wang's group has shown the possibility by generating AC voltage of up to 100 mV from the flexible SWG attached to the running hamster.

3. Transparent and Flexible Devices: Some of the piezoelectric nanostructure can be formed in various kinds of substrates, such as flexible and transparent organic substrate. The research groups in SKKU (Professor Sang-Woo Kim's group) and SAIT (Dr. Jae-Young Choi's group) have developed the transparent and flexible nanogenerator which can be possibly used for self-powered tactile sensor and anticipated that the development may be extended to the energy-efficient touch screen devices. Their research focus is being extended to enhance the transparency of the device and the cost-effectiveness by substituting Indium-Tin-Oxide (ITO) electrode with a graphene layer.

4. Implantable Telemetric Energy Receiver: The nanogenerator based on ZnO nanowire can be applied for implantable devices since ZnO not only is bio-compatible but also can be synthesized upon the organic substrate, rendering the nanogenerator bio-compatible in overall. The implantable device integrated with the nanogenerator can be operated by receiving the external ultrasonic vibration outside the human body, which is converted to the electrical energy by the piezoelectric nanostructure.

Triboelectric Nanogenerator

A triboelectric nanogenerator is an energy harvesting device that converts the external mechanical energy into electricity by a conjunction of triboelectric effect and electrostatic induction. This new type of nanogenerator was firstly demonstrated in Prof. Zhong Lin Wang's group at Georgia Institute of Technology in the year of 2012. As for this power generation unit, in the inner circuit, a potential is created by the triboelectric effect due to the charge transfer between two thin organic/inorganic films that exhibit opposite tribo-polarity; in the outer circuit, electrons are driven to flow between two electrodes attached on the back sides of the films in order to balance the potential.

Since the most useful materials for TENG are organic, it is also named organic nanogenerator, which is the first of using organic materials for harvesting mechanical energy.

Ever since the first report of the TENG in January 2012, the output power density of TENG has been improved for five orders of magnitude within 12 months. The area power density reaches 313 W/m², volume density reaches 490 kW/m³, and conversion efficiencies of ~60%–72% have been demonstrated. Besides the unprecedented output performance, this new energy technology also has a number of other advantages, such as low cost in manufacturing and fabrication, excellent robustness and reliability, and environmental friendliness. The triboelectric nanogenerator can be applied to harvest all kind mechanical energy that is available but wasted in our daily life, such as human motion, walking, vibration, mechanical triggering, rotating tire, wind, flowing water and more.

More importantly, Ramakrishna Podila's group at Clemson University demonstrated the first truly wireless triboelectric nanogenerators, which were able to wirelessly charge energy storage devices (e.g., batteries and capacitors) without need for any external amplification and boosters. These wireless generators could possibly pave the way for new systems that could be used to harvest mechanical energy and wirelessly transmit the generated energy for storage.

The triboelectric nanogenerator has three basic operation modes: vertical contact-separation mode, in-plane sliding mode, and single-electrode mode. They have different characteristics and are suitable for different applications.

Basic Modes and Mechanisms

Vertical Contact-separation Mode

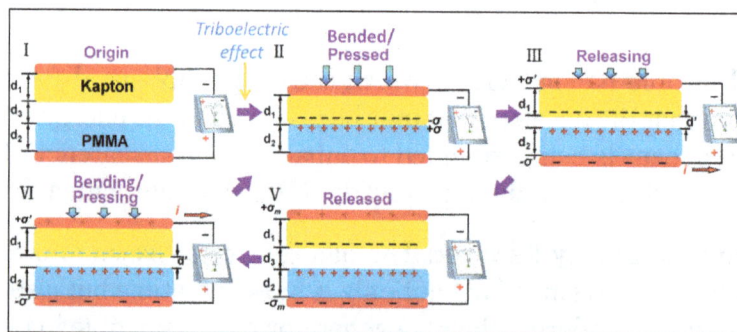

Vertical contact-separation mode of triboelectric nanogenerator.

The working mechanism of the triboelectric nanogenerator can be described as the periodic change of the potential difference induced by the cycled separation and re-contact of the opposite triboelectric charges on the inner surfaces of the two sheets. When a mechanical agitation is applied onto the device to bend or press it, the inners surfaces of the two sheets will get into close contact and the charge transfer will begin, leaving one side of the surface with positive charges and the other with negative charges. This is just the triboelectric effect. When the deformation is released, the two surfaces with opposite charges will separate automatically, so that these opposite triboelectric charges will generate an electric field in between and thus induce a potential difference across the top and bottom electrodes. In order to screen this potential difference, the electrons will be driven to flow from one electrode to the other through the external load. The electricity generated in this process will continue until the potentials of the two electrodes get back to even again. Subsequently, when the two sheets are pressed towards each other again, the triboelectric-charge-induced potential difference will begin to decrease to zero, so that the transferred charges will flow back through the external load, to generate another current pulse in the opposite direction. When this periodic mechanical deformation lasts, the alternating current (AC) signals will be continuously generated.

As for the pair of materials getting in contact and generating triboelectric charges, at least one of them need to be an insulator, so that the triboelectric charges cannot be conducted away but will remain on the inner surface of the sheet. Then, these immobile triboelectric charges can induce AC electricity flow in the external load under the periodic distance change.

Lateral Sliding Mode

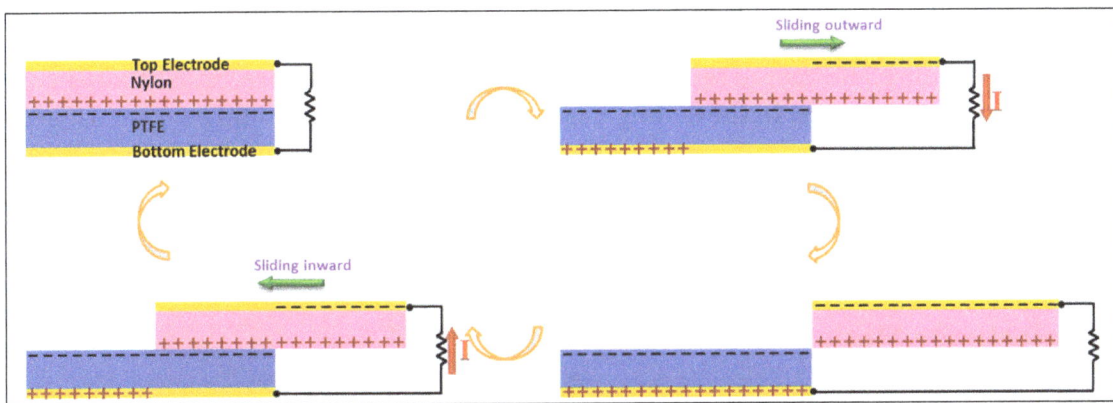

Lateral sliding mode of triboelectric nanogenerator.

There are two basic friction processes: normal contact, and lateral sliding. We demonstrated here a TENG that is designed based on the in-plane sliding between the two surfaces in lateral direction. With an intensive triboelectrification facilitated by sliding friction, a periodic change in the contact area between two surfaces leads to a lateral separation of the charge centers, which creates a voltage drop for driving the flow of electrons in the external load. The sliding-induced electricity generation mechanism is schematically depicted in the figure. In the original position, the two polymeric surfaces fully overlap and intimately contact with each other. Because of the large difference in the ability to attract electrons, the triboelectrification will leave one surface with net positive charges and the other with net negative charges with equal density. Since the tribo-charges on the insulators will only distribute in the surface layer and will not be leaked out for an extended period of time, the separation between the positively charged surface and negatively charged surface is negligible at this overlapping position, and thus there will be little electric potential drop across the two electrodes. Once the top plate with the positively charged surface starts to slide outward, the in-plane charge separation is initiated due to the decrease in contact surface area. The separated charges will generate an electric field pointing from the right to the left almost parallel to the plates, inducing a higher potential at the top electrode. This potential difference will drive a current flow from the top electrode to the bottom electrode in order to generate an electric potential drop that cancels the tribo-charge-induced potential. Because the vertical distance between the electrode layer and the tribo-charged polymeric surface is negligible compared to the lateral charge separation distance, the amount of the transferred charges on the electrodes approximately equals to the amount of the separated charges at any sliding displacement. Thus, the current flow will continue with the continuation of the ongoing sliding process that keeps increasing the separated charges, until the top plate fully slides out of the bottom plate and the tribo-charged surfaces are entirely separated. The measured current should be determined by the rate at which the two plates are being slid apart. Subsequently, when the top plate is reverted to slide backwards, the separated charges begins to get in contact again but no annihilation due to the insulator nature of the polymer materials. The redundant transferred charges on the electrodes will flow back through the external load with the increase of the contact area, in order to keep the electrostatic equilibrium. This will contribute to a current flow from the bottom electrode to the top electrode, along with the second half cycle of sliding. Once the two plates reach the overlapping position, the charged surfaces get into fully contact again. There will be no transferred charges left on the electrode, and the device returns to the first state. In this entire cycle, the processes of sliding outwards and inwards are symmetric, so a pair of symmetric alternating current peaks should be expected.

The mechanism of in-plane charge separation can work in either one directional sliding between two plates or in rotation mode. In the sliding mode, introducing linear grating or circular segmentation on the sliding surfaces is an extremely efficient means for energy harvesting. With such structures, two patterned triboelectric surfaces can get to fully mismatching position through a displacement of only a grating unit length rather than the entire length of the TENG so that it dramatically increase the transport efficiency of the induced charges.

Single-electrode Mode

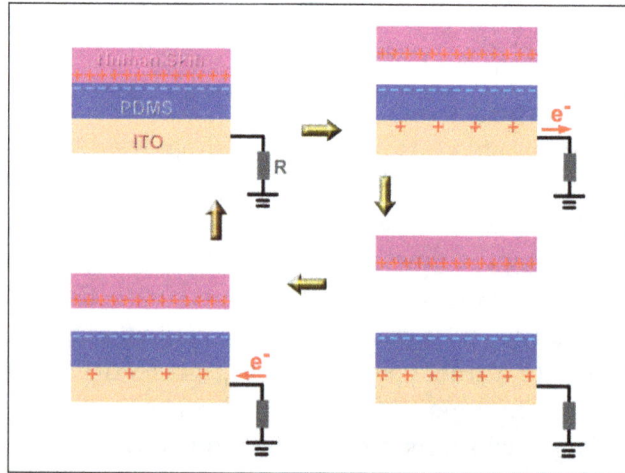

Single-electrode mode of triboelectric nanogenerator.

A single-electrode-based triboelectric nanogenerator is introduced as a more practical and feasible design for some applications such as fingertip-driven triboelectric nanoagenerator. The working principle of the single-electrode TENG is schematically shown in the figure by the coupling of contact electrification and electrostatic induction. In the original position, the surfaces of skin and PDMS fully contact with each other, resulting in charge transfer between them. According to the triboelectric series, electrons were injected from the skin to the PDMS since the PDMS is more triboelectrically negative than skin, which is the contact electrification process. The produced triboelectric charges with opposite polarities are fully balanced/screened, leading to no electron flow in the external circuit. Once a relative separation between PDMS and skin occurs, these triboelectric charges cannot be compensated. The negative charges on the surface of the PDMS can induce positive charges on the ITO electrode, driving free electrons to flow from the ITO electrode to ground. This electrostatic induction process can give an output voltage/current signal if the distance separating between the touching skin and the bottom PDMS is appreciably comparable to the size of the PDMS film. When negative triboelectric charges on the PDMS are fully screened from the induced positive charges on the ITO electrode by increasing the separation distance between the PDMS and skin, no output signals can be observed, as illustrated. Moreover, when the skin was reverted to approach the PDMS, the induced positive charges on the ITO electrode decrease and the electrons will flow from ground to the ITO electrode until the skin and PDMS fully contact with each other again, resulting in a reversed output voltage/current signal. This is a full cycle of electricity generation process for the TENG in contact-separation mode.

Applications

TENG is a physical process of converting mechanical agitation to an electric signal through the triboelectrification (in inner circuit) and electrostatic induction processes (in outer circuit). This basic process has been demonstrated for two major applications. The first application is energy harvesting with a particular advantage of harvesting mechanical energy. The other application is to serve as a self-powered active sensor, because it does not need an external power source to drive.

Harvesting Vibration Energy

Vibrations are a result of the most popular phenomena in society, from walking, voices, engine vibration, automobile, train, aircraft, wind and many more. It exists almost everywhere and at all the time. Harvesting vibration energy is of great value especially for powering mobile electronics, particularly in combination to complementary balanced energy harvesting techniques. Various technologies based on the fundamental principles of triboelectric nanogenerators have been demonstrated for harvesting vibration energy. This application of triboelectric nanogenerator has been demonstrated in the following aspects: 1. Cantilever-based technique is a classical approach for harvesting mechanical energy, especially for MEMS. By designing the contact surface of a cantilever with the top and bottom surfaces during vibration, TENG has been demonstrated for harvesting ambient vibration energy based on the contact-separation mode. 2. To harvest the energy from a backpack, we demonstrated a rationally designed TENG with integrated rhombic gridding, which greatly improved the total current output owing to the structurally multiplied unit cells connected in parallel. 3. With the use of 4 supporting springs, a harmonic resonator-based TENG has been fabricated based on the resonance induced contact-separation between the two triboelectric materials, which has been used to harvest vibration energy from an automobile engin, a sofa and a desk. 4. Recently, a three-dimensional triboelectric nanogenerator (3D-TENG) has been designed based on a hybridization mode of conjunction the vertical contact-separation mode and the in-plane sliding mode. The innovative design facilitates harvesting random vibration energy in multiple directions over a wide bandwidth. The 3-D TENG is designed for harvesting ambient vibration energy, especially at low frequencies, under a range of conditions in daily life, thus, opening the applications of TENG in environmental/infrastructure monitoring, charging portable electronics and internet of things.

Harvesting Energy from Human Body Motion

Since there is abundant mechanical energy generated on human bodies in people's everyday life, we can make use of the triboelectric nanogenerator to convert this amount of mechanical energy into electricity, for charging portable electronics and biomedical applications. This will help to greatly improve the convenience of people's life and expand the application of the personal electronics. A packaged power-generating insole with built-in flexible multi-layered triboelectric nanogenerators has been demonstrated, which enable harvesting mechanical pressure during normal walking. The TENG used here relies on the contact-separation mode and is effective in responding to the periodic compression of the insole. Using the insole as a direct power source, we develop a fully packaged self-lighting shoe that has broad applications for display and entertainment purposes. A TENG can be attached to the inner layer of a shirt for harvesting energy from body motion. Under the generally walking, the maximum output of voltage and current density are up to 17 V and 0.02 $\mu A/cm^2$, respectively. The TENG with a single layer size of 2 cm × 7 cm × 0.08 cm sticking on the clothes was demonstrated as a sustainable power source that not only can directly light up 30 light-emitting diodes (LEDs), but also can charge a lithium ion battery by persistently clapping clothes.

Self-powered Active Strain/Force Sensors

A triboelectric nanogenerator automatically generates an output voltage and current once it is mechanically triggered. The magnitude or the output signal signifies the impact of the mechanical deformation and its time-dependent behavior. This is the basic principle of the TENG can be applied as a self-powered pressure sensor. The voltage-output signal can reflect the applied pressure induced by a droplet of water. All types of TENGs have a high sensitivity and fast response to the external force and show as a sharp peak signal. Furthermore, the response to the impact of a piece of feather (20 mg, ~0.4 Pa in contact pressure) can be detected. The sensor signal can delicately show these details of the entire process. The existing results show that our sensor can be applied for measuring the subtle pressure in real life.

The active pressure sensor has also been developed in the form of a composite. The term of Triboelectric Composite refers to a sponge-shape polymer with embedded wire. Applying pressure and impact on the composite in any direction causes charge separation between the soft polymer and the active wire because of the presence of composite air gap. Passive wire as the second electrode may be either embedded inside the sponge without any air gap or placed out of the composite allowing the sensor to work in single electrode mode.

In a case that we make a matric array of the triboelectric nanogenerators, a large-area, and self-powered pressure map applied on a surface can be realized. The response of the TENG array with local pressure was measured through a multi-channel measurement system. There are two types of output signals from the TENG: open circuit voltage and short circuit current. The Open circuit voltage is only dictated by the final configuration of the TENG after applying a mechanical triggering, so that it is a measure of the magnitude of the deformation, which is attributed to the static information to be provided by TENG. The output current depends on the rate at which the induced charge would flow, so that the current signal is more sensitive to the dynamic process of how the mechanical triggering is applied.

The active pressure sensor and the integrated sensor array based on the triboelectric effect have several advantages over conventional passive pressure sensors. First, the active sensor is capable of both static pressure sensing using the open-circuit voltage and dynamic pressure sensing using the short-circuit current, while conventional sensors are usually incapable of dynamic sensing to provide the loading rate information. Second, the prompt response of both static and dynamic sensing enables the revealing of details about the loading pressure. Third, the detection limit of the TENG for dynamic sensing is as low as 2.1 Pa, owing to the high output of the TENG. Fourth, the active sensor array presented in this work has no power consumption and could even be combined with its energy harvesting functionality for self-powered pressure mapping. Future works in this field involve the miniaturization of the pixel size to achieve higher spatial resolution, and the integration of the TEAS matrix onto fully flexible substrate for shape-adaptive pressure imaging.

Self-powered Motion Sensors

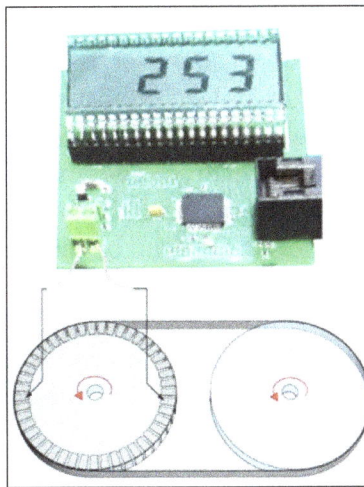

Smart belt-pulley system powers the encoder circuit by converting friction into electrical energy.

The term of self-powered sensors may reflect far beyond simple voltage-output signal. It can refer to a system which powers all the electronics responsible for measuring and demonstrating the detectable movement. For example, the self-powered triboelectric encoder, integrated in smart belt-pulley system, converts friction into useful electrical energy by storing the harvested energy in a capacitor and fully powering the circuit, including a microcontroller and an LCD.

Self-powered Active Chemical Sensors

As for triboelectric nanogenerators, maximizing the charge generation on opposite sides can be achieved by selecting the materials with the largest difference in the ability to attract electrons and changing the surface morphology. In such a case, the output of the TENG depends on the type and concentration of molecules adsorbed on the surface of the triboelectric materials, which can be used for fabricating chemical and biochemical sensors. As an example, the performance of the TENG depends on the assembly of Au nanoparticles (NPs) onto the metal plate. These assembled Au NPs not only act as steady gaps between the two plates at strain free condition, but also enable the function of enlarging the contact area of the two plates, which will increase the electrical output of the TENG. Through further modification of 3-mercaptopropionic acid (3-MPA) molecules on the assembled Au NPs, the high-output nanogenerator can become a highly sensitive and selective nanosensor toward Hg^{2+} ions detection because of the different triboelectric polarity of Au NPs and Hg^{2+} ions. With its high sensitivity, selectivity and simplicity, the TENG holds great potential for the determination of Hg^{2+} ions in environmental samples. The TENG is a future sensing system for unreachable and access-denied extreme environments. As different ions, molecules, and materials have their unique triboelectric polarities, we expect that the TENG can become either an electrical turn-on or turn-off sensor when the analytes are selectively binding to the modified electrode surface. We believe this work will serve as the stepping stone for related TENG studies and inspire the development of TENG toward other metal ions and biomolecules such as DNA and proteins in the near future.

Choice of Materials and Surface Structures

Almost all materials known exhibit the triboelectrification effect, from metal, to polymer, to silk and to wood, almost everything. All of these materials can be candidates for fabricating TENGs, so that the materials choices for TENG are huge. However, the ability of a material for gaining/losing electron depends on its polarity. John Carl Wilcke published the first triboelectric series in a 1757 on static charges. A material towards the bottom of the series, when touched to a material near the top of the series, will attain a more negative charge. The further away two materials are from each other on the series, the greater the charge transferred. Beside the choice of the materials in the triboelectric series, the morphologies of the surfaces can be modified by physical techniques with the creation of pyramids-, square- or hemisphere-based micro- or nano-patterns, which are effective for enhancing the contact area and possibly the triboelectrification. However, the created bumpy structure on the surface may increase the friction force, which may possibly reduce the energy conversion efficiency of the TENG. Therefore, an optimization has to be designed for maximizing the conversion efficiency.

The surfaces of the materials can be functionalized chemically using various molecules, nanotubes, nanowires or nanoparticles, in order to enhance the triboelectrification effect. Surface functionalization can largely change the surface potential. The introduction of nanostructures on the surfaces can change the local contact characteristics, which may improve the triboelectrification. This will involve a large amount of studies for testing a range of materials and a range of available nanostructures.

Besides these pure materials, the contact materials can be made of composites, such embedding nanoparticles in polymer matrix. This not only changes the surface electrification, but also the permittivity of the materials so that they can be effective for electrostatic induction. Therefore, there are numerous ways for enhancing the performance of the TENG from the materials point of view. This gives an excellent opportunity for chemists and materials scientists to do extensive study both in the basic science and in practical application. In contrast, materials systems for solar cell and thermal electric, for example, are rather limited, and there are not very many choices for high performance devices.

Pyroelectric Nanogenerator

A pyroelectric nanogenerator is an energy harvesting device converting the external thermal energy into an electrical energy by using nano-structured pyroelectric materials. Usually, harvesting thermoelectric energy mainly relies on the Seebeck effect that utilizes a temperature difference between two ends of the device for driving the diffusion of charge carriers. However, in an environment that the temperature is spatially uniform without a gradient, such as in outdoor in our daily life, the Seebeck effect can not be used to harvest thermal energy from a time-dependent temperature fluctuation. In this case, the pyroelectric effect has to be the choice, which is about the spontaneous polarization in certain anisotropic solids as a result of temperature fluctuation. The first pyroelectric nanogenerator was introduced by Prof. Zhong Lin Wang at Georgia Institute of Technology in 2012. By harvesting the waste heat energy, this new type of nanogenerator has the potential applications such as wireless sensors, temperature imaging, medical diagnostics, and personal electronics.

Mechanism

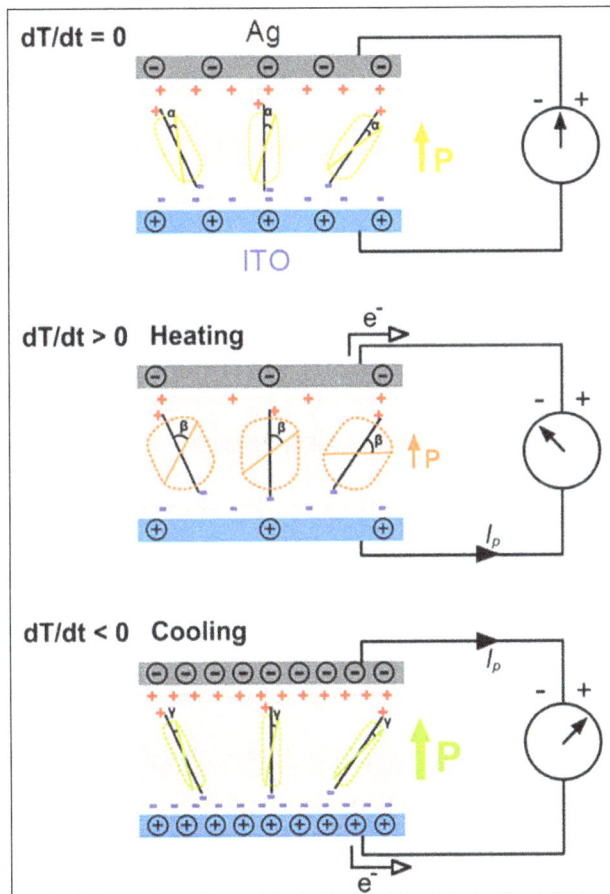

In figure, the mechanism of the pyroelectric nanogenerator based on a composite structure of pyroelectric nanowries. (a-c) Schematic diagrams of the pyroelectric nanogenerator with negative electric dipoles under room temperature (a), heated (b) and cooled (c) conditions. The angles marked in the diagrams represent the degrees to which the dipole would oscillate as driven by statistical thermal fluctuations.

The working principle of pyroelectric nanogenerator will be explained for 2 different cases: the primary pyroelectric effect and the secondary pyroelectric effect.

The working principle for the first case is explained by the primary pyroelectric effect, which describes the charge produced in a strain-free case. The primary pyroelectric effect dominates the pyroelectric response in PZT, BTO, and some other ferroelectric materials. The mechanism is based on the thermally induced random wobbling of the electric dipole around its equilibrium axis, the magnitude of which increases with increasing temperature. Due to thermal fluctuations under room temperature, the electric dipoles will randomly oscillate within a degree from their respective aligning axes. Under a fixed temperature, the total average strength of the spontaneous polarization form the electric dipoles is constant, resulting in no output of the pyroelectric nanogenerator. If we apply a change in temperature in the nanogenerator from room temperature to a higher temperature, the increase in temperature will result in that the electric dipoles oscillate within a larger degree of spread around their respective aligning axes. The total average spontaneous polarization is decreased due to the spread of the oscillation angles. The quantity of induced charges in the electrodes are thus reduced, resulting in a flow of electrons. If the nanogenerator is cooled instead of heated, the spontaneous polarization will be enhanced since the electric dipoles oscillate within a smaller degree of spread angles due to the lower thermal activity. The total magnitude of the polarization is increased and the amount of induced charges in the electrodes are increased. The electrons will then flow in an opposite direction.

For the second case, the obtained pyroelectric response is explained by the secondary pyroelectric effect, which describes the charge produced by the strain induced by thermal expansion. The secondary pyroelectric effect dominates the pyroelectric response in ZnO, CdS, and some other wurzite-type materials. The thermal deformation can induce a piezoelectric potential difference across the material, which can drive the electrons to flow in the external circuit. The output of the nanogenerator is associated with the piezoelectric coefficient and the thermal deformation of the materials. The output current I of the pyroelectric nanogenerators can be determined by the equation of $I = pA(dT/dt)$, where p is the pyroelectric coefficient, A is the effective area of the NG, dT/dt is the rate of change in temperature.

Applications

Pyroelectric nanogenerator is expected to be applied for various applications where the time-dependent temperature fluctuation exists. One of the feasible applications of the pyroelectric nanogenerator is used as an active sensor, which can work without a battery. One example has been introduced by Professor Zhong Lin Wang's group in 2012 by using a pyroelectric nanogenerator as the self-powered temperature sensor for detecting a change in temperature, where the response time and reset time of the sensor are about 0.9 and 3 s, respectively. In general, the pyroelectric nanogenerator gives a high output voltage, but the output current is small. It not only can be used as a potential power source, but also as an active sensor for measuring temperature variation.

NANOSENSOR

Nanosensors are nanoscale devices that measure physical .quantities and convert those quantities to signals that can be detected and analyzed. There are several ways being proposed today to make nanosensors; these include top-down lithography, bottom-up assembly, and molecular self-assembly.

There are different types of nanosensors in the market and in development for various applications. Though all sensors measure different things, sensors share the same basic workflow: a selective binding of an analyte, signal generation from the interaction of the nanosensor with the bio-element, and processing of the signal into useful metrics.

Characteristics

Nanomaterials-based sensors have several benefits in sensitivity and specificity over sensors made from traditional materials. Nanosensors can have increased specificity because they operate at a similar scale as natural biological processes, allowing functionalization with chemical and biological molecules, with recognition events that cause detectable physical changes. Enhancements in sensitivity stem from the high surface-to-volume ratio of nanomaterials, as well as novel physical properties of nanomaterials that can be used as the basis for detection, including nanophotonics. Nanosensors can also potentially be integrated with nanoelectronics to add native processing capability to the nanosensor.

In addition to their sensitivity and specificity, nanosensors offer significant advantages in cost and response times, which makes nanosensors suitable for high-throughput applications. Nanosensors provide real-time monitoring compared to traditional detection methods such as chromatography and spectroscopy. These traditional methods may take days to weeks to obtain results and often require investment in capital costs as well as time for sample preparation.

One-dimensional nanomaterials such as nanowires and nanotubes are well suited for use in nanosensors, as compared to bulk or thin-film planar devices. They can function both as transducers and wires to transmit the signal. Their high surface area can cause large signal changes upon binding of an analyte. Their small size can enable extensive multiplexing of individually addressable sensor units in a small device. Their operation is also "label free" in the sense of not requiring fluorescent or radioactive labels on the analytes.

There are several challenges for nanosensors, including avoiding fouling and drift, developing reproducible calibration methods, applying preconcentration and separation methods to attain a proper analyte concentration that avoids saturation, and integrating the nanosensor with other elements of a sensor package in a reliable manufacturable manner. Because nanosensors are a relatively new technology, there are many unanswered questions regarding nanotoxicology which currently limits their application in biological systems. Some nanosensors may impact cell metabolism and homeostasis, changing cellular molecular profiles and making it difficult to separate sensor-induced artifacts from fundamental biological phenomena.

Potential applications for nanosensors include medicine, detection of contaminants and pathogens, and monitoring manufacturing processes and transportation systems. By measuring changes in physical properties (volume, concentration, displacement and velocity, gravitational, electrical, and magnetic forces, pressure, or temperature) nanosensors may be able to distinguish between and recognize certain cells at the molecular level in order to deliver medicine or monitor development to specific places in the body. The type of signal transduction defines the major classification system for nanosensors. Some of the main types of nanosensor readouts include optical, mechanical, vibrational, or electromagnetic.

Binding	Signal Generation	Processing
Analyte binds selectively to the nanosensor.	Based on the interaction of the sensor with the bio-element, a signal is produced.	The signal can then be processed into useful metrics.

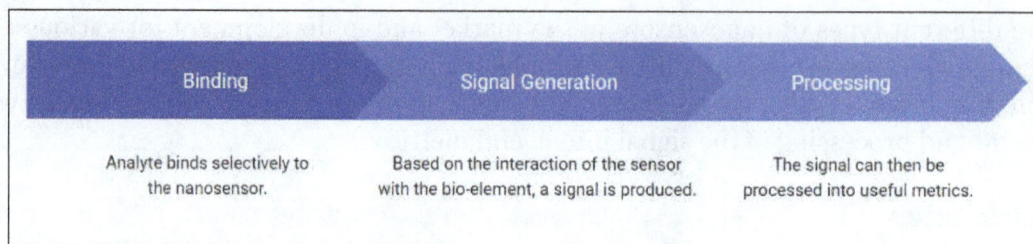

Overview of a general nanosensor workflow.

Mechanisms of Operation

There are many mechanisms by which a recognition event can be transduced into a measurable signal. Electrochemical nanosensors are based on detecting a resistance change in the nanomaterial upon binding of an analyte, due to changes in scattering or to the depletion or accumulation of charge carriers. One possibility is to use nanowires such as carbon nanotubes, conductive polymers, or metal oxide nanowires as gates in field-effect transistors, although as of 2009 they had not yet been demonstrated in real-world conditions. Chemical nanosensors contain a chemical recognition system (receptor) and a physiochemical transducer, in which the receptor interacts with analyte to produce electrical signals. Other examples include electromagnetic or plasmonic nanosensors, spectroscopic nanosensors such as surface-enhanced Raman spectroscopy, magnetoelectronic or spintronic nanosensors, and mechanical nanosensors.

Photonic devices can also be used as nanosensors to quantify concentrations of clinically relevant samples. A principle of operation of these sensors is based on the chemical modulation of a hydrogel film volume that incorporates a Bragg grating. As the hydrogel swells or shrinks upon chemical stimulation, the Bragg grating changes color and diffracts light at different wavelengths. The diffracted light can be correlated with the concentration of a target analyte.

Production Methods

There are currently several hypothesized ways to produce nanosensors. Top-down lithography is the manner in which most integrated circuits are now made. It involves starting out with a larger block of some material and carving out the desired form. These carved out devices, notably put to use in specific microelectromechanical systems used as microsensors, generally only reach the micro size, but the most recent of these have begun to incorporate nanosized components.

Another way to produce nanosensors is through the bottom-up method, which involves assembling the sensors out of even more minuscule components, most likely individual atoms or molecules. This would involve moving atoms of a particular substance one by one into particular positions which, though it has been achieved in laboratory tests using tools such as atomic force microscopes, is still a significant difficulty, especially to do en masse, both for logistic reasons as well as economic ones. Most likely, this process would be used mainly for building starter molecules for self-assembling sensors.

The third way, which promises far faster results, involves self-assembly, or "growing" particular nanostructures to be used as sensors. This most often entails an already complete set of components that would automatically assemble themselves into a finished product. Accurately being able to reproduce this effect for a desired sensor in a laboratory would imply that scientists could manufacture nanosensors much more quickly and potentially far more cheaply by letting numerous molecules assemble themselves with little or no outside influence, rather than having to manually assemble each sensor.

Applications

One of the first working examples of a synthetic nanosensor was built by researchers at the Georgia Institute of Technology in 1999. It involved attaching a single particle onto the end of a carbon nanotube and measuring the vibrational frequency of the nanotube both with and without the particle. The discrepancy between the two frequencies allowed the researchers to measure the mass of the attached particle.

Since then, increasing amounts of research have gone into nanosensors, whereby modern nanosensors have been developed for many applications. Currently, the applications of nanosensors in the market include: healthcare, defense and military, and others such as food, environment, and agriculture.

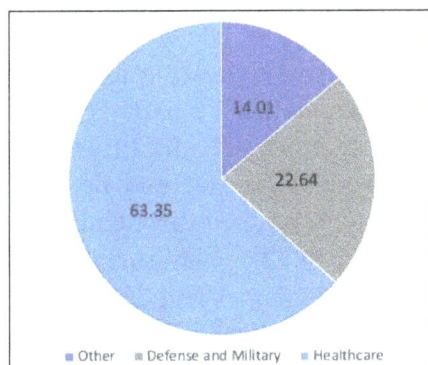

Brief breakdown of current industry applications of nanosensors.

Defense and Military

Nanoscience as a whole has much potential for applications in the defense and military sector. Applications include chemical detection, decontamination, and forensics. However, the primary efforts surrounding nanosensors largely remain in research and development.

Some nanosensors in development for defense applications include nanosensors for the detection of explosives or toxic gases. Such nanosensors work on the principle that gas molecules can be distinguished based on their mass using, for example, piezoelectric sensors. If a gas molecule is adsorbed at the surface of the detector, the resonance frequency of the crystal changes and this can be measured as a change in electrical properties. In addition, field effect transistors, used as potentiometers, can detect toxic gases if their gate is made sensitive to them.

In a similar application, nanosensors can be utilized in military and law enforcement clothing and gear. The Navy Research Laboratory's Institute for Nanoscience has studied quantum dots for application in nanophotonics and identifying biological materials. Nanoparticles layered with polymers and other receptor molecules will change color when contacted by analytes such as toxic gases. This alerts the user that they are in danger. Other projects involve embedding clothing with biometric sensors to relay information regarding the user's health and vitals, which would be useful for monitoring soldiers in combat.

Surprisingly, some of the most challenging aspects in creating nanosensors for defense and military use are political in nature, rather than technical. Many different government agencies must work together to allocate budgets and share information and progress in testing; this can be difficult with such large and complex institutions. In addition, visas and immigration status can become an issue for foreign researchers - as the subject matter is very sensitive, government clearance can sometimes be required. Finally, there are currently not well defined or clear regulations on nanosensor testing or applications in the sensor industry, which contributes to the difficulty of implementation.

Food and the Environment

Nanosensors can improve various sub-areas within food and environment sectors including food processing, agriculture, air and water quality monitoring, and packaging and transport. These nanosensors enable rapid analysis of samples and detection of contaminants in food which is especially important for downstream processing and perishable items.

Chemical sensors are useful for analyzing odors from food samples and detecting atmospheric gases. The "electronic nose" was developed in 1988 to determine the quality and freshness of food samples using traditional sensors, but more recently the sensing film has been improved with nanomaterials. A sample is placed in a chamber where volatile compounds become concentrated in the gas phase, whereby the gas is then pumped through the chamber to carry the aroma to the sensor that measures its unique fingerprint. The high surface area to volume ratio of the nanomaterials allows for greater interaction with analytes and the nanosensor's fast response time enables the separation of interfering responses. Chemical sensors, too, have been built using nanotubes to detect various properties of gaseous molecules. Carbon nanotubes have been used to sense ionization of gaseous molecules while nanotubes made out of titanium have been employed to detect atmospheric concentrations of hydrogen at the molecular level. Many of these involve a system by which nanosensors are built to have a specific pocket for another molecule. When that particular molecule, and only that specific molecule, fits into the nanosensor, and light is shone upon the nanosensor, it will reflect different wavelengths of light and, thus, be a different color. In a similar fashion, Flood et al. have shown that supramolecular host–guest chemistry offers quantitative sensing using Raman scattered light as well as SERS.

Other types of nanosensors, including quantum dots and gold nanoparticles, are currently being developed to detect pollutants and toxins in the environment. Quantum dot surfaces can be modified with antibodies that bind specifically to environmental pollutants. Gold nanoparticle-based optical sensors can be used to detect heavy metals very precisely; for example, mercury levels as low as 0.49 nM. This sensing modality takes advantage of fluorescence resonance energy transfer (FRET), in which the presence of metals inhibits the interaction between quantum dots and gold nanoparticles, and quenches the FRET response.

The main challenge associated with using nanosensors in food and the environment is determining their associated toxicity and overall effect on the environment. Currently, there is insufficient knowledge on how the implementation of nanosensors will affect the soil, plants, and humans in the long-term. This is difficult to fully address because nanoparticle toxicity depends heavily on the type, size, and dosage of the particle as well as environmental variables including pH, temperature, and humidity To mitigate potential risk, research is being done to manufacture safe, nontoxic nanomaterials, as part of an overall effort towards green nanotechnology.

Healthcare

One example of nanosensors involves using the fluorescence properties of cadmium selenide quantum dots as sensors to uncover tumors within the body. A downside to the cadmium selenide dots, however, is that they are highly toxic to the body. As a result, researchers are working on developing alternate dots made out of a different, less toxic material while still retaining some of the fluorescence properties. In particular, they have been investigating the particular benefits of zinc sulfide quantum dots which, though they are not quite as fluorescent as cadmium selenide, can be augmented with other metals including manganese and various lanthanide elements. In addition, these newer quantum dots become more fluorescent when they bond to their target cells.

Another application of nanosensors involves using silicon nanowires in IV lines to monitor organ health. The nanowires are sensitive to detect trace biomarkers that diffuse into the IV line through blood which can monitor kidney or organ failure. These nanowires would allow for continuous biomarker measurement, which provides some benefits in terms of temporal sensitivity over traditional biomarker quantification assays such as ELISA.

Nanosensors can also be used to detect contamination in organ implants. The nanosensor is embedded into the implant and detects contamination in the cells surrounding the implant through an electric signal sent to a clinician or healthcare provider. The nanosensor can detect whether the cells are healthy, inflammatory, of contaminated with bacteria.

Currently, there are stringent regulations in place for the development of standards for nanosensors to be used in the medical industry, due to insufficient knowledge of the adverse effects of nanosensors as well as potential cytotoxic effects of nanosensors. Additionally, there can be a high cost of raw materials such as silicon, nanowires, and carbon nanotubes, which prevent commercialization and manufacturing of nanosensors requiring scale-up for implementation. To mitigate the drawback of cost, researchers are looking into manufacturing nanosensors made of more cost-effective materials. There is also a high degree of precision needed to reproducibly manufacture nanosensors, due to their small size and sensitivity to different synthesis techniques, which creates additional technical challenges to be overcome.

NANOELECTROMECHANICAL SYSTEMS

Nanoelectromechanical systems (NEMS) are a class of devices integrating electrical and mechanical functionality on the nanoscale. NEMS form the logical next miniaturization step from so-called microelectromechanical systems, or MEMS devices. NEMS typically integrate transistor-like nanoelectronics with mechanical actuators, pumps, or motors, and may thereby form physical, biological, and chemical sensors. The name derives from typical device dimensions in the nanometer range, leading to low mass, high mechanical resonance frequencies, potentially large quantum mechanical effects such as zero point motion, and a high surface-to-volume ratio useful for surface-based sensing mechanisms. Uses include accelerometers, or detectors of chemical substances in the air.

As noted by Richard Feynman in his famous talk in 1959, "There's Plenty of Room at the Bottom," there are many potential applications of machines at smaller and smaller sizes; by building and controlling devices at smaller scales, all technology benefits. Among the expected benefits include greater efficiencies and reduced size, decreased power consumption and lower costs of production in electromechanical systems.

In 2000, the first very-large-scale integration (VLSI) NEMS device was demonstrated by researchers at IBM. Its premise was an array of AFM tips which can heat/sense a deformable substrate in order to function as a memory device. Further devices have been described by Stefan de Haan. In 2007, the International Technical Roadmap for Semiconductors (ITRS) contains NEMS Memory as a new entry for the Emerging Research Devices section.

Atomic Force Microscopy

A key application of NEMS is atomic force microscope tips. The increased sensitivity achieved by NEMS leads to smaller and more efficient sensors to detect stresses, vibrations, forces at the atomic level, and chemical signals. AFM tips and other detection at the nanoscale rely heavily on NEMS.

Approaches to Miniaturization

Two complementary approaches to fabrication of NEMS can be found. The top-down approach uses the traditional microfabrication methods, i.e. optical, electron beam lithography and thermal treatments, to manufacture devices. While being limited by the resolution of these methods, it allows a large degree of control over the resulting structures. In this manner devices such as nanowires, nanorods, and patterned nanostructures are fabricated from metallic thin films or etched semiconductor layers.

Bottom-up approaches, in contrast, use the chemical properties of single molecules to cause single-molecule components to self-organize or self-assemble into some useful conformation, or rely on positional assembly. These approaches utilize the concepts of molecular self-assembly and molecular recognition. This allows fabrication of much smaller structures, albeit often at the cost of limited control of the fabrication process.

A combination of these approaches may also be used, in which nanoscale molecules are integrated into a top-down framework. One such example is the carbon nanotube nanomotor.

Materials

Carbon Allotropes

Many of the commonly used materials for NEMS technology have been carbon based, specifically diamond, carbon nanotubes and graphene. This is mainly because of the useful properties of carbon based materials which directly meet the needs of NEMS. The mechanical properties of carbon (such as large Young's modulus) are fundamental to the stability of NEMS while the metallic and semiconductor conductivities of carbon based materials allow them to function as transistors.

Both graphene and diamond exhibit high Young's modulus, low density, low friction, exceedingly low mechanical dissipation, and large surface area. The low friction of CNTs, allow practically frictionless bearings and has thus been a huge motivation towards practical applications of CNTs as constitutive elements in NEMS, such as nanomotors, switches, and high-frequency oscillators. Carbon nanotubes and graphene's physical strength allows carbon based materials to meet higher stress demands, when common materials would normally fail and thus further support their use as a major materials in NEMS technological development.

Along with the mechanical benefits of carbon based materials, the electrical properties of carbon nanotubes and graphene allow it to be used in many electrical components of NEMS. Nanotransistors have been developed for both carbon nanotubes as well as graphene. Transistors are one of the basic building blocks for all electronic devices, so by effectively developing usable transistors, carbon nanotubes and graphene are both very crucial to NEMS.

Metallic Carbon Nanotubes

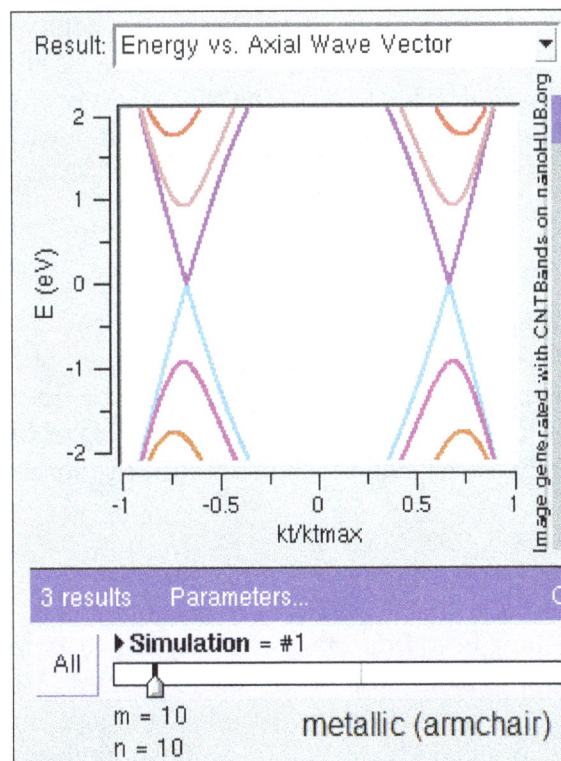

Band structures computed using tight binding approximation for (6,0) CNT (zigzag, metallic), (10,2) CNT (semiconducting) and (10,10) CNT (armchair, metallic).

Carbon nanotubes (CNTs) are allotropes of carbon with a cylindrical nanostructure. They can be considered a rolled up graphene. When rolled at specific and discrete ("chiral") angles, and the combination of the rolling angle and radius decides whether the nanotube has a bandgap (semi-conducting) or no bandgap (metallic).

Metallic carbon nanotubes have also been proposed for nanoelectronic interconnects since they can carry high current densities. This is a useful property as wires to transfer current are another basic building block of any electrical system. Carbon nanotubes have specifically found so much use in NEMS that methods have already been discovered to connect suspended carbon nanotubes to other nanostructures. This allows carbon nanotubes to form complicated nanoelectric systems. Because carbon based products can be properly controlled and act as interconnects as well as transistors, they serve as a fundamental material in the electrical components of NEMS.

Difficulties

Despite all of the useful properties of carbon nanotubes and graphene for NEMS technology, both of these products face several hindrances to their implementation. One of the main problems is carbon's response to real life environments. Carbon nanotubes exhibit a large change in electronic properties when exposed to oxygen. Similarly, other changes to the electronic and mechanical attributes of carbon based materials must fully be explored before their implementation, especially because of their high surface area which can easily react with surrounding environments. Carbon nanotubes were also found to have varying conductivities, being either metallic or semiconducting depending on their helicity when processed. Because of this, special treatment must be given to the nanotubes during processing to assure that all of the nanotubes have appropriate conductivities. Graphene also has complicated electric conductivity properties compared to traditional semiconductors because it lacks an energy band gap and essentially changes all the rules for how electrons move through a graphene based device. This means that traditional constructions of electronic devices will likely not work and completely new architectures must be designed for these new electronic devices.

Biohybrid NEMS

The emerging field of bio-hybrid systems combines biological and synthetic structural elements for biomedical or robotic applications. The constituting elements of bio-nanoelectromechanical systems (BioNEMS) are of nanoscale size, for example DNA, proteins or nanostructured mechanical parts. Examples include the facile top-down nanostructuring of thiol-ene polymers to create cross-linked and mechanically robust nanostructures that are subsequently functionalized with proteins.

Simulations

Computer simulations have long been important counterparts to experimental studies of NEMS devices. Through continuum mechanics and molecular dynamics (MD), important behaviors of NEMS devices can be predicted via computational modeling before engaging in experiments. Additionally, combining continuum and MD techniques enables engineers to efficiently analyze the stability of NEMS devices without resorting to ultra-fine meshes and time-intensive simulations. Simulations have other advantages as well: they do not require the time and expertise associated with fabricating NEMS devices; they can effectively predict the interrelated roles of various electromechanical effects; and parametric studies can be conducted fairly readily as compared with experimental approaches. For example, computational studies have predicted the charge distributions and "pull-in" electromechanical responses of NEMS devices. Using simulations to predict mechanical and electrical behavior of these devices can help optimize NEMS device design parameters.

CARBON NANOTUBES IN PHOTOVOLTAICS

Organic photovoltaic devices (OPVs) are fabricated from thin films of organic semiconductors, such as polymers and small-molecule compounds, and are typically on the order of 100 nm thick. Because polymer based OPVs can be made using a coating process such as spin coating or inkjet printing, they are an attractive option for inexpensively covering large areas as well as flexible plastic surfaces. A promising low cost alternative to conventional solar cells made of crystalline silicon, there is a large amount of research being dedicated throughout industry and academia towards developing OPVs and increasing their power conversion efficiency.

Single Wall Carbon Nanotubes as Light Harvesting Media

Single wall carbon nanotubes possess a wide range of direct bandgaps matching the solar spectrum, strong photoabsorption, from infrared to ultraviolet, and high carrier mobility and reduced carrier transport scattering, which make themselves ideal photovoltaic material. Photovoltaic effect can be achieved in ideal single wall carbon nanotube (SWNT) diodes. Individual SWNTs can form ideal p-n junction diodes. An ideal behavior is the theoretical limit of performance for any diode, a highly sought after goal in all electronic materials development. Under illumination, SWNT diodes show significant power conversion efficiencies owing to enhanced properties of an ideal diode.

Recently, SWNTs were directly configured as energy conversion materials to fabricate thin-film solar cells, with nanotubes serving as both photogeneration sites and a charge carriers collecting/transport layer. The solar cells consist of a semitransparent thin film of nanotubes conformally coated on a n-type crystalline silicon substrate to create high-density p-n heterojunctions between nanotubes and n-Si to favor charge separation and extract electrons (through n-Si) and holes (through nanotubes). Initial tests have shown a power conversion efficiency of >1%, proving that CNTs-on-Si is a potentially suitable configuration for making solar cells. For the first time, Zhongrui Li demonstrated that SOCl2 treatment of SWNT boosts the power conversion efficiency of SWNT/n-Si heterojunction solar cells by more than 60%. Later on the acid doping approach is widely adopted in the later published CNT/Si works. Even higher efficiency can be achieved if acid liquid is kept inside the void space of nanotube network. Acid infiltration of nanotube networks significantly boosts the cell efficiency to 13.8%,as reported by Yi Jia, by reducing the internal resistance that improves fill factor, and by forming photoelectrochemical units that enhance charge separation and transport. The wet acid induced problems can be avoided by using aligned CNT film. In aligned CNT film, the transport distance is shortened, and the exciton quenching rate is also reduced. Additionally aligned nanotube film has much smaller void space, and better contact with substrate. So, plus strong acid doping, using aligned single wall carbon nanotube film can further improve power conversion efficiency (a record-high power-conversion-efficiency of >11% was achieved by Yeonwoong Jung).

Zhongrui Li also made the first n-SWNT/p-Si photovoltaic device by tuning SWNTs from p-type to n-type through polyethylene imine functionalization.

Carbon Nanotube Composites in the Photoactive Layer

Combining the physical and chemical characteristics of conjugated polymers with the high conductivity along the tube axis of carbon nanotubes (CNTs) provides a great deal of incentive to disperse CNTs into the photoactive layer in order to obtain more efficient OPV devices. The interpenetrating bulk donor–acceptor heterojunction in these devices can achieve charge separation and collection because of the existence of a bicontinuous network. Along this network, electrons and holes can travel toward their respective contacts through the electron acceptor and the polymer hole donor. Photovoltaic efficiency enhancement is proposed to be due to the introduction of internal polymer/nanotube junctions within the polymer matrix. The high electric field at these junctions can split up the excitons, while the single-walled carbon nanotube (SWCNT) can act as a pathway for the electrons.

The dispersion of CNTs in a solution of an electron donating conjugated polymer is perhaps the most common strategy to implement CNT materials into OPVs. Generally poly(3-hexylthiophene) (P3HT) or poly(3-octylthiophene) (P3OT) are used for this purpose. These blends are then spin coated onto a transparent conductive electrode with thicknesses that vary from 60 to 120 nm. These conductive electrodes are usually glass covered with indium tin oxide (ITO) and a 40 nm sublayer of poly(3,4-ethylenedioxythiophene) (PEDOT) and poly(styrenesulfonate) (PSS). PEDOT and PSS help to smooth the ITO surface, decreasing the density of pinholes and stifling current leakage that occurs along shunting paths. Through thermal evaporation or sputter coating, a 20 to 70 nm thick layer of aluminum and sometimes an intermediate layer of lithium fluoride are then applied onto the photoactive material. Multiple research investigations with both multi-walled carbon nanotubes (MWCNTs) and single-walled carbon nanotubes (SWCNTs) integrated into the photoactive material have been completed.

Enhancements of more than two orders of magnitude have been observed in the photocurrent from adding SWCNTs to the P3OT matrix. Improvements were speculated to be due to charge separation at polymer–SWCNT connections and more efficient electron transport through the SWCNTs. However, a rather low power conversion efficiency of 0.04% under 100 mW/cm² white illumination was observed for the device suggesting incomplete exciton dissociation at low CNT concentrations of 1.0% wt. Because the lengths of the SWCNTs were similar to the thickness of photovoltaic films, doping a higher percentage of SWCNTs into the polymer matrix was believed to cause short circuits. To supply additional dissociation sites, other researchers have physically blended functionalized MWCNTs into P3HT polymer to create a P3HT-MWCNT with fullerene C_{60} double-layered device. However, the power efficiency was still relatively low at 0.01% under 100 mW/cm² white illumination. Weak exciton diffusion toward the donor–acceptor interface in the bilayer structure may have been the cause in addition to the fullerene C_{60} layer possibly experiencing poor electron transport.

More recently, a polymer photovoltaic device from C_{60} modified SWCNTs and P3HT has been fabricated. Microwave irradiating a mixture of aqueous SWCNT solution and C_{60} solution in toluene was the first step in making these polymer-SWCNT composites. Conjugated polymer P3HT was then added resulting in a power conversion efficiency of 0.57% under simulated solar irradiation (95 mW/cm²). It was concluded that improved short circuit current density was a direct result of the addition of SWCNTs into the composite causing faster electron transport via the network of SWCNTs. It was also concluded that the morphology change led to an improved fill factor. Overall, the main result was improved power conversion efficiency with the addition of SWCNTs, compared to cells without SWCNTs; however, further optimization was thought to be possible.

Additionally, it has been found that heating to the point beyond the glass transition temperature of either P3HT or P3OT after construction can be beneficial for manipulating the phase separation of the blend. This heating also affects the ordering of the polymeric chains because the polymers are microcrystalline systems and it improves charge transfer, charge transport, and charge collection throughout the OPV device. The hole mobility and power efficiency of the polymer-CNT device also increased significantly as a result of this ordering.

Emerging as another valuable approach for deposition, the use of tetraoctylammonium bromide in tetrahydrofuran has also been the subject of investigation to assist in suspension by exposing SWCNTs to an electrophoretic field. In fact, photoconversion efficiencies of 1.5% and 1.3% were achieved when SWCNTs were deposited in combination with light harvesting cadmium sulfide (CdS) quantum dots and porphyrins, respectively.

Among the best power conversions achieved to date using CNTs were obtained by depositing a SWCNT layer between the ITO and the PEDOT : PSS or between the PEDOT : PSS and the photo-active blend in a modified ITO/PEDOT : PSS/ P3HT : (6,6)-phenyl-C_{61}-butyric acid methyl ester (PCBM)/Al solar cell. By dip-coating from a hydrophilic suspension, SWCNT were deposited after an initially exposing the surface to an argon plasma to achieve a power conversion efficiency of 4.9%, compared to 4% without CNTs.

However, even though CNTs have shown potential in the photoactive layer, they have not resulted in a solar cell with a power conversion efficiency greater than the best tandem organic cells (6.5% efficiency). But, it has been shown in most of the previous investigations that the control over a uniform blending of the electron donating conjugated polymer and the electron accepting CNT is one of the most difficult as well as crucial aspects in creating efficient photocurrent collection in CNT-based OPV devices. Therefore, using CNTs in the photoactive layer of OPV devices is still in the initial research stages and there is still room for novel methods to better take advantage of the beneficial properties of CNTs.

One issue with utilizing SWCNTs for the photoactive layer of PV devices is the mixed purity when synthesized (about 1/3 metallic and 2/3 semiconducting). Metallic SWCNTs (m-SWCNTs) can cause exciton recombination between the electron and hole pairs, and the junction between metallic and semiconducting SWCNTs (s-SWCNTs) form Schottky barriers that reduce the hole transmission probability. The discrepancy in electronic structure of synthesized CNTs requires electronic sorting to separate and remove the m-SWCNTs to optimize the semiconducting performance. This may be accomplished through diameter and electronic sorting of CNTs through a density gradient ultracentrifugation (DGU) process, involving a gradient of surfactants that can separate the CNTs by diameter, chirality, and electronic type. This sorting method enables the separation of m-SWCNTs and the precise collection of multiple chiralities of s-SWCNTs, each chirality able to absorb a unique wavelength of light. The multiple chiralities of s-SWCNTs are used as the hole transport material along with the fullerene component PC71BM to fabricate heterojunctions for the PV active layer. The polychiral s-SWCNTs enable a wide range optical absorption from visible to near-infrared (NIR) light, increasing the photo current relative to using single chirality nanotubes. To maximize light absorption, the inverted device structure was used with a zinc oxide nanowire layer penetrating the active layer to minimize collection length. Molybdenum oxide (MoOx) was utilized as a high work function hole transport layer to maximize voltage.

Cells fabricated with this architecture have achieved record power conversion efficiencies of 3.1%, higher than any other solar cell materials that utilize CNTs in the active layer. This design also has exceptional stability, with the PCE remaining at around 90% over a period of 30 days. The exceptional chemical stability of carbon nanomaterials enables excellent environmental stability compared to most organic photovoltaics that must be encapsulated to reduce degradation.

Relative to the best of polymer-fullerene heterojunction solar cells that have PCEs of about 10%, polychiral nanotube and fullerene solar cells are still a far ways off. Nevertheless, these findings push the achievable limits of CNT technology in solar cells. The ability for polychiral nanotubes to absorb in the NIR regime is a technology that can be utilized to improve the efficiencies of future of multi-junction tandem solar cells along with increasing the lifetime and durability of future noncrystalline solar cells.

Carbon Nanotubes as a Transparent Electrode

ITO is currently the most popular material used for the transparent electrodes in OPV devices; however, it has a number of deficiencies. For one, it is not very compatible with polymeric substrates due to its high deposition temperature of around 600 °C. Traditional ITO also has unfavorable mechanical properties such as being relatively fragile. In addition, the combination of costly layer deposition in vacuum and a limited supply of indium results in high quality ITO transparent electrodes being very expensive. Therefore, developing and commercializing a replacement for ITO is a major focus of OPV research and development.

Conductive CNT coatings have recently become a prospective substitute based on wide range of methods including spraying, spin coating, casting, layer-by-layer, and Langmuir–Blodgett deposition. The transfer from a filter membrane to the transparent support using a solvent or in the form of an adhesive film is another method for attaining flexible and optically transparent CNT films. Other research efforts have shown that films made of arc-discharge CNT can result in a high conductivity and transparency. Furthermore, the work function of SWCNT networks is in the 4.8 to 4.9 eV range (compared to ITO which has a lower work function of 4.7 eV) leading to the expectation that the SWCNT work function should be high enough to assure efficient hole collection. Another benefit is that SWCNT films exhibit a high optical transparency in a broad spectral range from the UV-visible to the near-infrared range. Only a few materials retain reasonable transparency in the infrared spectrum while maintaining transparency in the visible part of the spectrum as well as acceptable overall electrical conductivity. SWCNT films are highly flexible, do not creep, do not crack after bending, theoretically have high thermal conductivities to tolerate heat dissipation, and have high radiation resistance. However, the electrical sheet resistance of ITO is an order of magnitude less than the sheet resistance measured for SWCNT films. Nonetheless, initial research studies demonstrate SWCNT thin films can be used as conducting, transparent electrodes for hole collection in OPV devices with efficiencies between 1% and 2.5% confirming that they are comparable to devices fabricated using ITO. Thus, possibilities exist for advancing this research to develop CNT-based transparent electrodes that exceed the performance of traditional ITO materials.

CNTs in Dye-sensitized Solar Cells

Due to the simple fabrication process, low production cost, and high efficiency, there is significant interest in dye-sensitized solar cells (DSSCs). Thus, improving DSSC efficiency has been the subject of a variety of research investigations because it has the potential to be manufactured economically enough to compete with other solar cell technologies. Titanium dioxide nanoparticles have been widely used as a working electrode for DSSCs because they provide a high efficiency, more than any other metal oxide semiconductor investigated. Yet the highest conversion efficiency under air mass (AM) 1.5 (100 mW/cm^2) irradiation reported for this device to date is about 11%. Despite this initial success, the effort to further enhance efficiency has not produced any major results. The transport of electrons across the particle network has been a key problem in achieving higher photoconversion efficiency in nanostructured electrodes. Because electrons encounter many grain boundaries during the transit and experience a random path, the probability of their recombination with oxidized sensitizer is increased. Therefore, it is not adequate to enlarge the oxide electrode surface area to increase efficiency because photo-generated charge recombination should be prevented. Promoting electron transfer through film electrodes and blocking interface states lying below the edge of the conduction band are some of the non-CNT based strategies to enhance efficiency that have been employed.

With recent progress in CNT development and fabrication, there is promise to use various CNT based nanocomposites and nanostructures to direct the flow of photogenerated electrons and assist in charge injection and extraction. To assist the electron transport to the collecting electrode surface in a DSSC, a popular concept is to utilize CNT networks as support to anchor light harvesting semiconductor particles. Research efforts along these lines include organizing CdS quantum dots on SWCNTs. Charge injection from excited CdS into SWCNTs was documented upon excitation of CdS nanoparticles. Other varieties of semiconductor particles including CdSe and CdTe can induce charge-transfer processes under visible light irradiation when attached to CNTs. Including porphyrin and C$_{60}$ fullerene, organization of photoactive donor polymer and acceptor fullerene on electrode surfaces has also been shown to offer considerable improvement in the photoconversion efficiency of solar cells. Therefore, there is an opportunity to facilitate electron transport and increase the photoconversion efficiency of DSSCs utilizing the electron-accepting ability of semiconducting SWCNTs.

Other researchers fabricated DSSCs using the sol-gel method to obtain titanium dioxide coated MWCNTs for use as an electrode. Because pristine MWCNTs have a hydrophobic surface and poor dispersion stability, pretreatment was necessary for this application. A relatively low-destruction method for removing impurities, H_2O_2 treatment was used to generate carboxylic acid groups by oxidation of MWCNTs. Another positive aspect was the fact that the reaction gases including CO_2 and H_2O were non-toxic and could be released safely during the oxidation process. As a result of treatment, H_2O_2 exposed MWCNTs have a hydrophilic surface and the carboxylic acid groups on the surface have polar covalent bonding. Also, the negatively charged surface of the MWCNTs improved the stability of dispersion. By then entirely surrounding the MWCNTs with titanium dioxide nanoparticles using the sol-gel method, an increase in the conversion efficiency of about

50% compared to a conventional titanium dioxide cell was achieved. The enhanced interconnectivity between the titanium dioxide particles and the MWCNTs in the porous titanium dioxide film was concluded to be the cause of the improvement in short circuit current density. Here again, the addition of MWCNTs was thought to provide more efficient electron transfer through film in the DSSC.

One issue with utilizing SWCNTs for the photoactive layer of PV devices is the mixed purity when synthesized (about 1/3 metallic and 2/3 semiconducting). Metallic SWCNTs (m-SWCNTs) can cause exciton recombination between the electron and hole pairs, and the junction between metallic and semiconducting SWCNTs (s-SWCNTs) form Schottky barriers that reduce the hole transmission probability. The discrepancy in electronic structure of synthesized CNTs requires electronic sorting to separate and remove the m-SWCNTs to optimize the semiconducting performance. This may be accomplished through diameter and electronic sorting of CNTs through a density gradient ultracentrifugation (DGU) process, involving a gradient of surfactants that can separate the CNTs by diameter, chirality, and electronic type. This sorting method enables the separation of m-SWCNTs and the precise collection of multiple chiralities of s-SWCNTs, each chirality able to absorb a unique wavelength of light. The multiple chiralities of s-SWCNTs are used as the hole transport material along with the fullerene component PC71BM to fabricate heterojunctions for the PV active layer. The polychiral s-SWCNTs enable a wide range optical absorption from visible to near-infrared (NIR) light, increasing the photo current relative to using single chirality nanotubes. To maximize light absorption, the inverted device structure was used with a zinc oxide nanowire layer penetrating the active layer to minimize collection length. Molybdenum oxide (MoOx) was utilized as a high work function hole transport layer to maximize voltage.

Cells fabricated with this architecture have achieved record power conversion efficiencies of 3.1%, higher than any other solar cell materials that utilize CNTs in the active layer. This design also has exceptionally stability, with the PCE remaining at around 90% over a period of 30 days. The exceptional chemical stability of carbon nanomaterials enables excellent environmental stability compared to most organic photovoltaics that must be encapsulated to reduce degradation.

Relative to the best of polymer-fullerene heterojunction solar cells that have PCEs of about 10%, polychiral nanotube and fullerene solar cells are still a far ways off. Nevertheless, these findings push the achievable limits of CNT technology in solar cells. The ability for polychiral nanotubes to absorb in the NIR regime is a technology that can be utilized to improve the efficiencies of future of multi-junction tandem solar cells along with increasing the lifetime and durability of future noncrystalline solar cells.

CARBON NANOTUBE NANOMOTOR

A device generating linear or rotational motion using carbon nanotube as the primary component, is termed a nanotube nanomotor. Nature already has some of the most efficient and powerful kinds of nanomotors. Some of these natural biological nanomotors have been re-engineered to serve desired purposes. However, such biological nanomotors are designed to work in specific environmental conditions (pH, liquid medium, sources of energy, etc.). Laboratory-made nanotube nanomotors on the other hand are significantly more robust and can operate in diverse environments including varied frequency, temperature, mediums and chemical environments. The vast differences in the dominant forces and criteria between macroscale and micro/nanoscale offer new avenues to construct tailor-made nanomotors. The various beneficial properties of carbon nanotubes makes them the most attractive material to base such nanomotors on.

Size Effects

Electrostatic Forces

Coulomb's law states that the electrostatic force between two objects is inversely proportional to the square of their distance. Hence, as the distance is reduced to less than a few micrometers, a large force can be generated from seemingly small charges on two bodies. However, electrostatic charge scales quadratically, thereby the electrostatic force also scales quadratically, as the following equations show:

$$C = \frac{\varepsilon A}{d} \propto L$$

$$E \propto L^0$$

$$V = E * L = E * L \propto L$$

$$Q = CV \propto L^2$$

$$F = A * E^2 \propto L^2$$

Alternatively,

$$F = \frac{Q_1 * Q_2}{d^2} \propto \frac{L^2 * L^2}{L^2} \propto L^2$$

Here, A is area, C is capacitance, F is electrostatic force, E is electrostatic field, L is length, V is voltage and Q is charge. Despite the scaling nature of the electrostatic force it is one of the major mechanisms of sensing and actuation in the field of microelectromechanical systems (MEMS) and is the backbone for the working mechanism of the first NEMS nanomotor. The quadratic scaling is alleviated by increasing the number of units generating the electrostatic force as seen in comb drives in many MEMS devices.

Friction

Just as the electrostatic force, the frictional force scales quadratically with size $F \sim L^2$.

Friction is an ever plaguing problem regardless of the scale of a device. It becomes all the more prominent when a device is scaled down. In the nano scale it can wreak havoc if not accounted for because the parts of a Nano-Electro-Mechanical-Systems (NEMS) device are sometimes only a few atoms thick. Furthermore, such NEMS devices typically have a very large surface area-to-volume ratio. Surfaces in the nanoscale resemble a mountain range, where each peak corresponds to an atom or a molecule. Friction at the nanoscale is proportional to the number of atoms that interact between two surfaces. Hence, friction between perfectly smooth surfaces in the macroscale is actually similar to large rough objects rubbing against each other.

In the case of nanotube nanomotors however, the intershell friction in the multi-walled nanotubes (MWNT) is remarkably small. Molecular dynamics studies show that, with the exception of small peaks, the frictional force remains almost negligible for all sliding velocities until a special sliding velocity is reached. Simulations relating the sliding velocity, induced rotation, inter-shell frictional force to the applied force provide explanations for the low inter-wall friction. Contrary to macroscale expectations the speed at which an inner tube travels within an outer tube does not follow a linear relationship with the applied force. Instead, the speed remains constant (as in a plateau) despite increasing applied force occasionally jumping in value to the next plateau. No real rotation is noticed in nonchiral inner tubes. In the case of chiral tubes a true rotation is noticed and the angular velocity also jumps to plateaus along with the jumps in the linear velocity. These plateaus and jumps can be explained as a natural outcome of frictional peaks for growing velocity, the stable (rising) side of the peak leading to a plateau, the dropping (unstable) side leading to a jump. These peaks occur due to parametric excitation of vibrational modes in the walls of the tubes due to the sliding of the inner tube. With the exception of small peaks, that correspond to the speed plateaus, the frictional force remains almost negligible for all sliding velocities until a special sliding velocity. These velocity plateaus correspond to the peaks in the frictional force. The sudden rise in sliding velocity is due to a resonance condition between a frequency that is dependent on the inter-tube corrugation period and particular phonon frequencies of the outer tube which happen to possess a group velocity approximately equal to the sliding velocity.

First NEMS Nanomotor

The first nanomotor can be thought of as a scaled down version of a comparable microelectromechanical systems (MEMS) motor. The nanoactuator consists of a gold plate rotor, rotating about the axis of a multi-walled nanotube (MWNT). The ends of the MWNT rest on a SiO_2 layer which form the two electrodes at the contact points. Three fixed stator electrodes (two visible 'in-plane' stators and one 'gate' stator buried beneath the surface) surround the rotor assembly. Four independent voltage signals (one to the rotor and one to each stators) are applied to control the position, velocity and direction of rotation. Empirical angular velocities recorded provide a lower bound of 17 Hz (although capable of operating at much higher frequencies) during complete rotations.

Fabrication

The MWNTs are synthesized by the arc-discharge technique, suspended in 1,2-dichlorobenzene and deposited on degenerately doped silicon substrates with a 1 μm of SiO_2. The MWNT can be aligned according to pre-made markings on the substrate by using an atomic force microscope (AFM) or a scanning electron microscope (SEM). The rotor, electrodes and the 'in-plane' stators are patterned using electron beam lithography using an appropriately masked photo-resist. Gold with a chromium adhesion layer is thermally evaporated, lifted off in acetone and then annealed at 400 °C to ensure better electrical and mechanical contact with the MWNT. The rotor measures 250–500 nm on a side. An HF etch is then used to remove sufficient thickness (500 nm of SiO_2) of the substrate to make room for the rotor when it rotates. The Si substrate serves as the gate stator. The MWNT at this point displays a very high torsional spring constant (10^{-15} to 10^{-13} N m with resonant frequencies in the tens of megahertz), hence, preventing large angular displacements. To overcome this, one or more outer MWNT shells are compromised or removed in the region between the anchors and the rotor plate. One simple way to accomplish this is by successively applying very large stator voltages (around 80 V DC) that cause mechanical fatigue and eventually shear the outer shells of the MWNT. An alternative method involves the reduction of the outermost MWNT tubes to smaller, wider concentric nanotubes beneath the rotor plate.

The smaller nanotubes are fabricated using the Electrical driven vaporization (EDV) which is a variant of the electrical-breakdown technique. Passing current between the two electrodes typically results in failure of the outermost shell only on one side of the nanotube. Current is therefore passed between one electrode and the center of the MWNT which results in the failure of the outermost shell between this electrode and the center. The process is repeated on the opposite side to result in the formation of the short concentric nanotube that behaves like a low friction bearing along the longer tube.

Arrays of Nanoactuators

Due to the miniscular magnitude of output generated by a single nanoactuator the necessity to use arrays of such actuators to accomplish a higher task comes into picture. Conventional methods like chemical vapor deposition (CVD) allow the exact placement of nanotubes by growing them directly on the substrate. However, such methods are unable to produce very high qualities of MWNT. Moreover, CVD is a high temperature process that would severely limit the compatibility with other materials in the system. A Si substrate is coated with electron beam resist and soaked in acetone to leave only a thin polymer layer. The substrate is selectively exposed to an low energy electron beam of an SEM that activates the adhesive properties of the polymer later. This forms the basis for the targeting method. The alignment method exploits the surface velocity obtained by a fluid as it flows off a spinning substrate. MWNTs are suspended in orthodicholrobenzene (ODCB) by ultrasonication in an aquasonic bath that separates most MWNT bundles into individual MWNTs. Drops of this suspension are then pipetted one by one onto the center of a silicon substrate mounted on a spin coater rotating at 3000 rpm. Each subsequent drop of the suspension is pipetted only after the previous drop has completely dried to ensure larger density and better alignment of the MWNTs (90% of the MWNTs over 1 µm long lie within 1°). Standard electron beam lithography is used to pattern the remaining components of the nanoactuators.

Arc-discharge Evaporation Technique

This technique is a variant of the standard arc-discharge technique used for the synthesis of fullerenes in an inert gas atmosphere. As Figure shows, the experiment is carried out in a reaction vessel containing an inert gas such as helium, argon, etc. flowing at a constant pressure. A potential of around 18 V is applied across two graphite electrodes (diameters of the anode and cathode are 6 mm and 9 mm) separated by a short distance of usually 1–4 mm within this chamber. The amount of current (usually 50–100 A) passed through the electrodes to ensure nanotube formation depends on the dimensions of the electrodes, separation distance and the inert gas used. As a result, carbon atoms are ejected from the anode and are deposited onto the cathode hence shrinking the mass of the anode and increasing the mass of the cathode. The black carbonaceous deposit (a mixture of nanoparticles and nanotubes in a ratio of 1:2) is seen growing on the inside of the cathode while a hard grey metallic shell forms on the outside. The total yield of nanotubes as a proportion of starting graphitic material peaks at a pressure of 500 torr at which point 75% of graphite rod consumed is converted to nanotubes. The nanotubes formed range from 2 to 20 nm in diameter and few to several micrometers in length. There are several advantages of choosing this method over the other techniques such as laser ablation and chemical vapor deposition such as fewer structural defects (due to high growth temperature), better electrical, mechanical and thermal properties, high production rates (several hundred mg in ten minutes), etc.

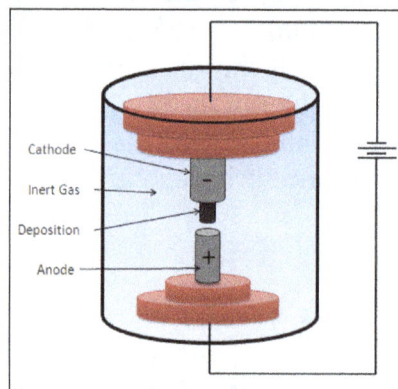

The basic experimental setup for the arc-discharge technique of large scale carbon nanotube synthesis.

Electrical-breakdown Technique

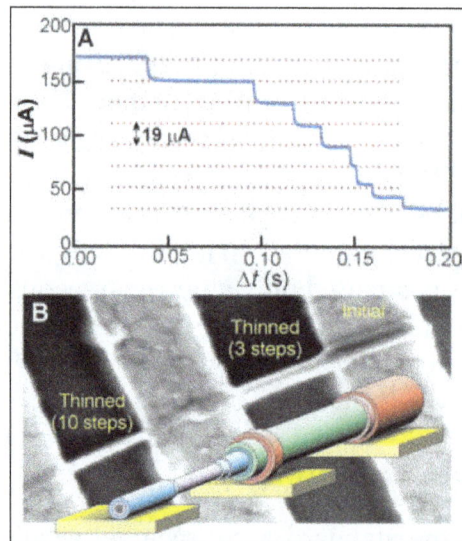

In figure, (A) Graph showing remarkably discrete, constant drops in conductance for the removal of each subsequent carbon shell under constant voltage (B) Images of partially broken MWNTs show clear thinning, with a decrease in radius equal to the intershell spacing (0.34 nm) times the number of completed breakdown steps. The two segments of this sample were independently thinned by 3 and 10 shells, as depicted by the color overlays.

Large-scale synthesis of carbon nanotubes typically results in a randomly varied proportion of different types of carbon nanotubes. Some may be semiconducting while others may be metallic in their electrical properties. Most applications require the use of such specific types of nanotubes. Electrical-breakdown technique provides a means for separating and selecting desired type of nanotubes. Carbon nanotubes are known to withstand very large current densities up to 10^9 A/cm^2 partly due to the strong sigma bonds between carbon atoms. However, at sufficiently high currents the nanotubes fail primarily due to rapid oxidation of the outermost shell. This results in a partial conductance drop that becomes apparent within a few seconds. Applying an increased bias displays multiple independent and stepwise drops in conductance resulting from the sequential failure of carbon shells. Current in a MWNT typically travels in the outermost shell due the direct contact between this shell and the electrodes. This controlled destruction of shells without affecting disturbing inner layers of MWNTs permits the effective separation of the nanotubes.

Principle

The rotor is made to rotate using electrostatic actuation. An out-of-phase common frequency sinusoidal voltages to two in-plane stators S_1, S_2, a doubled frequency voltage signal to the gate stator S_3 and a DC offset voltage to the rotor plate R are applied as shown below:

$$S_1 = V_0 sin(\omega t)$$

$$S_2 = V_0 sin(\omega t - \pi)$$

$$S_3 = V_0 sin(2\omega t + \pi / 2)$$

$$R = -V_0$$

By the sequential application of these asymmetrical stator voltages (less than 5 V) the rotor plate can be drawn to successive stators hence making the plate complete rotations. The high proximity between the stators and the rotor plate is one reason why a large force is not required for electrostatic actuation. Reversing the bias causes the rotor to rotate in the opposite direction as expected.

Applications

- The rotating metal plate could serve as a mirror for ultra-high-density optical sweeping and switching devices as the plate is at the limit of visible light focusing. An array of such actuators, each serving as a high frequency mechanical filter, could be used for parallel signal processing in telecommunications.

- The plate could serve as a paddle for inducing or detecting fluid motion in microfluidic applications. It could serve as a bio-mechanical element in biological systems, a gated catalyst in wet chemistry reactions or as a general sensor element.

- A charged oscillating metal plate could be used as a transmitter of electromagnetic radiation.

Thermal Gradient Driven Nanotube Actuators

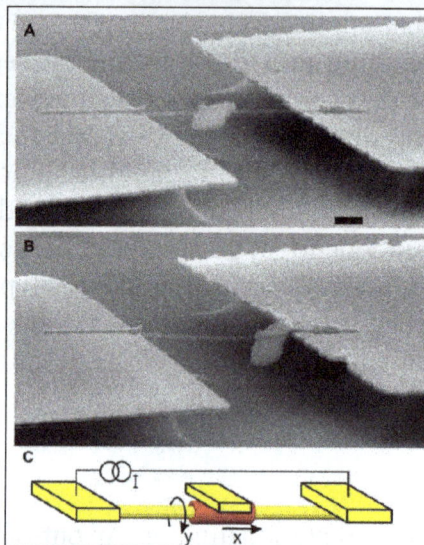

Thermal gradient driven nanomotor. (A & B): SEM images of experimental setup.
(C) Schematic of the nanomotor also displaying degrees of freedom.

The nanoactuator, comprises two electrodes connected via a long MWNT. A gold plate acts as the cargo and is attached to a shorter and wider concentric nanotube. The cargo moves towards the cooler electrode due to the thermal gradient in the longer nanotube induced by the high current that is passed through it. The maximum velocity was approximated to 1 µm/s which is comparable to the speeds attained by kinesin biomotors.

Fabrication

The MWNT are fabricated using the standard arc-discharge evaporation process and deposited on an oxidized silicon substrate. The gold plate in the center of the MWNT is patterned using electron-beam lithography and Cr/Au evaporation. During the same process, the electrodes are attached to the nanotube. Finally, electrical-breakdown technique is used to selectively remove a few outer walls of the MWNT. Just as the nanoactuator from the Zettl group, this enables low friction rotation and translation of the shorter nanotube along the axis of the longer tube. The application of the electrical-breakdown technique does not result in the removal of the tubes below the cargo. This might be because the metal cargo absorbs the heat generated in the portion of the tube in its immediate vicinity hence delaying or possibly even preventing tube oxidation in this part.

Principle

The interaction between the longer and shorter tubes generates an energy surface that confines the motion to specific tracks – translation and rotation. The degree of translational and rotational motion of the shorter tube are highly dependent on the chiralities of the two tubes. Motion in the nanoactuator displayed a proclivity of the shorter tube to follow a path of minimum energy. This path could either have a roughly constant energy or have a series of barriers. In the former case, friction and vibrational motion of atoms can be neglected whereas a stepwise motion is expected in the latter scenario.

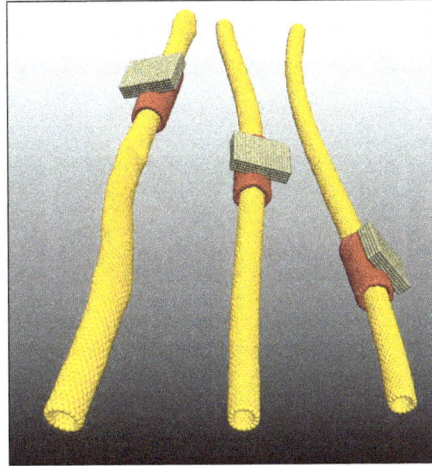

Motion of shorter nanotubes (red) along the longer tubes (yellow) from the hotter(top) section of the nanotube to the cooler (bottom) section of the nanotube carrying the metal cargo (gray).

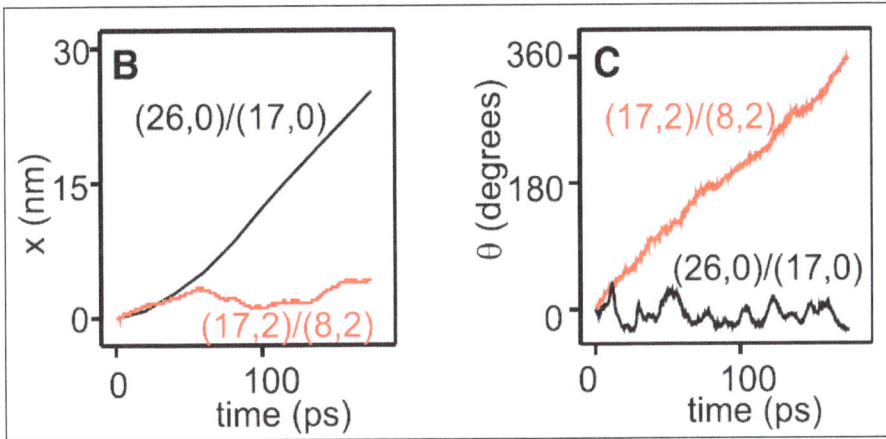

Degree of translational and rotation are dependent on the chiralities of the two nanotubes.

Stepwise Motion

The stepwise motion can be explained by the existence of periodic energy barriers for relative motion between the longer and shorter tubes. For a given pair of nanotubes, the ratio of the step in rotation to the step in translation is typically a constant, the value of which depends on the chirality of the nanotubes. The energy of such barriers could be estimated from the temperature in the nanotube, a lower bound for which can be estimated as the melting temperature of gold (1300 K) by noting that the gold plate melts to form a spherical structure as current is passed through the nanomotor. The motion rate γ can be written as a function of the attempt frequency ω, the Boltzmann's constant k, and temperature T as:

$$T = \frac{\omega}{2\pi} e^{\frac{-\Delta E}{kT}}$$

Taking $\Gamma \approx 1 Hz$, using the approximation:

$$\omega = \sqrt{\frac{\Delta E}{m a_0^2}}$$

where, m is the mass of the cargo and a_0^2 represents the contact area, the barrier height is estimated as 17 μeV per atom.

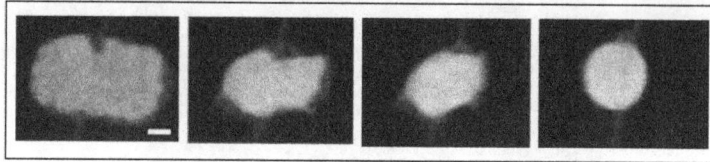

SEM images show the transformation of gold plate (left) into a ball (right) due to very high temperatures.

Mechanism for Actuation

Many proposals were made to explain the driving mechanism behind the nanoactuator. The high current (0.1 mA) required to drive the actuator is likely to cause sufficient dissipation to clean the surface of contaminants; hence, ruling out the possibility of contaminants playing a major role. The possibility of electromigration, where the electrons move atomic impurities via momentum transfer due to collisions, was also ruled out because the reversal of the current direction did not affect the direction of displacement. Similarly, rotational motion could not have been caused by an induced magnetic field due to the current passing through the nanotube because the rotation could either be left or right-handed depending on the device. Stray electric field effect could not be the driving factor because the metal plate staid immobile for high resistive devices even under a large applied potential. The thermal gradient in the nanotube provides the best explanation for the driving mechanism.

Thermal Gradient Induced Motion

The induced motion of the shorter nanotube is explained as the reverse of the heat dissipation that occurs in friction wherein the sliding of two objects in contact results in the dissipation of some of the kinetic energy as phononic excitations caused by the interface corrugation. The presence of a thermal gradient in a nanotube causes a net current of phononic excitations traveling from the hotter region to the cooler region. The interaction of these phononic excitations with mobile elements (the carbon atoms in the shorter nanotube) causes the motion of the shorter nanotube. This explains why the shorter nanotube moves towards the cooler electrode. Changing the direction of the current has no effect on the shape of thermal gradient in the longer nanotube. Hence, direction of the movement of the cargo is independent of the direction of the bias applied. The direct dependence of the velocity of the cargo to the temperature of the nanotube is inferred from the fact that the velocity of the cargo decreases exponentially as the distance from the midpoint of the long nanotube increases.

Shortcomings

The temperatures and the thermal gradient that the MWNT are subjected to are very high. On one hand, the high thermal gradient seems to have a highly detrimental effect on the lifetime of such nanoactuators. On the other hand, experiments show that the displacement of the shorter tube is directly proportional to the thermal gradient. Therefore, a compromise needs to be reached to optimize the thermal gradient. The dimensions of movable nanotube is directly related to the energy barrier height. Although the current model excites multiple phonon modes, selective phonon mode excitation would enable lowering the phonon bath temperature.

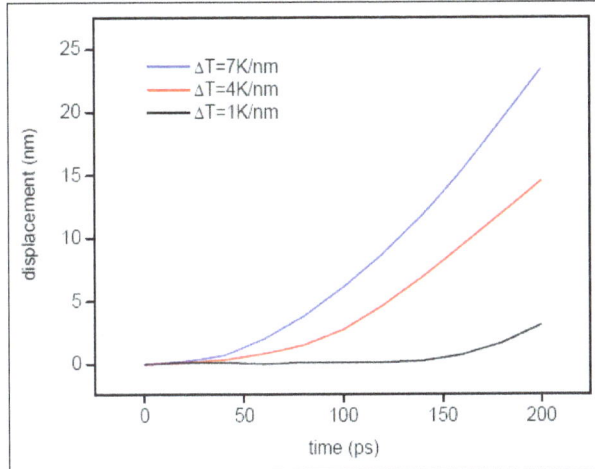

Graph demonstrating the direct relationship between the thermal gradient and the displacement of the shorter tube/cargo.

Applications

- Pharmaceutical/Nanofluidic: Thermal gradient could be used to drive fluids within the nanotubes or in nanofluidic devices as well as for drug delivery by nanosyringes.

- Running bio-engineered nanopores using heat generated from adenosine triphosphate (ATP) molecules.

Electron Windmill

Structure

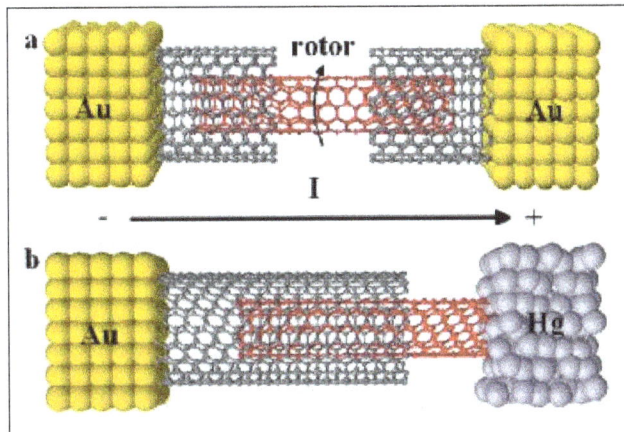

MWNT nanomotor (A) and nanodrill (B).

As figure shows, the nanomotor consists of a double-walled CNT (DWNT) formed from an achiral (18,0) outer tube clamped to external gold electrodes and a narrower chiral (6,4) inner tube. The central portion of the outer tube is removed using the electrical-breakdown technique to expose the free-to-rotate, inner tube. The nanodrill also comprises an achiral outer nanotube attached to a gold electrode but the inner tube is connected to a mercury bath.

Principle

Conventional nanotube nanomotors make use of static forces that include elastic, electrostatic, friction and van der Waals forces. The electron windmill model makes use of a new "electron-turbine" drive mechanism that obviates that need for metallic plates and gates that the above nano-actuators require. When a DC voltage is applied between the electrodes, a "wind" of electrons is produced from left to right. The incident electron flux in the outer achiral tube initially possesses zero angular momentum, but acquires a finite angular momentum after interacting with the inner chiral tube. By Newton's third law, this flux produces a tangential force (hence a torque) on the inner nanotube causing it to rotate hence giving this model the name – "electron windmill". For moderate voltages, the tangential force produced by the electron wind is much greatly exceed the associated frictional forces.

Applications

Some of the main applications of the electron windmill include:

- A voltage pulse could cause the inner element to rotate at a calculated angle hence making the device behave as a switch or a nanoscale memory element.

- Modification of the electron windmill to construct a nanofluidic pump by replacing the electrical contacts with reservoirs of atoms or molecules under the influence of an applied pressure difference.

NANOWIRE

Nanowire is a thin wire, generally having a diameter less than or equal to 100 nanometers (1 nm = 1×10^{-9} metre). The first nanoscale quantum-well wire (a thinly layered semiconductor structure) was developed in 1987 by scientists at Bell Laboratories. A nanowire of more-refined design was developed and described in 1991 by Belgium engineer Jean-Pierre Colinge. Since then, nanowires have been investigated for possible applications in many fields, including optics, electronics, and genetics.

Nanowires can be made from a wide variety of materials, including silicon, germanium, carbon, and various conductive metals, such as gold and copper. Their small size makes them good conductors, with electrons passing easily through them, a property that has allowed for important advances in computer science. For example, the development of an optical photonic switch using specialized cadmium sulfide nanowires that allow photons to pass through the wire and act as binary signals (i.e., 0 and 1) has the potential to greatly enhance computer speed.

In genetics, researchers have used nanowires to create artificial protein-coding DNA. Such nanowires are formed using amino acids, which are the building blocks of proteins, and DNA. The technology could be used to facilitate the creation or production of proteins, thereby advancing protein research and potentially leading to advances in therapeutic applications, such as for the replacement or repair of dysfunctional proteins.

References

- Nanoelectronics: circuitstoday.com, Retrieved 17 June, 2019

- Wang, Xufeng; Muralidharan, Bhaskaran; Klimeck, Gerhard (2006). "nanoHUB.org - Resources: Coulomb Blockade Simulation". nanoHUB. doi:10.4231/d3c24qp1w

- Kaufman, Igor Kh; McClintock, Peter V E; Eisenberg, Robert S (2015). "Coulomb blockade model of permeation and selectivity in biological ion channels". New Journal of Physics. 17 (8): 083021. Bibcode:2015N-JPh...17h3021K. doi:10.1088/1367-2630/17/8/083021

- Pal, Malay; Somalwar, Neha; Singh, Anumeha; Bhat, Ramray; Eswarappa, Sandeep; Saini, Deepak; Ghosh, Ambarish (2018). "Maneuverability of Magnetic Nanomotors Inside Living Cells". Advanced Materials. 30 (22): 1800429. doi:10.1002/adma.201800429. PMID 29635828

- Heggo, A (2013). "NanoBatteries Technology Application". International Journal on Power and Engineering and Energy. 4. doi:10.12986/IJPEE.2013.010

- Fennimore, A. M.; T.D. Yuzvinsky; Wei-Qiang Han; M. S. Fuhrer; J. Cumings; A. Zettl (2003). "Rotational actuators based on carbon nanotubes". Nature. 424 (6947): 408–410. Bibcode:2003Natur.424..408F. doi:10.1038/nature01823. PMID 12879064

- Nanowire, technology: britannica.com, Retrieved 20 May, 2019

Nanophotonics

The study of the behavior of light on the nanometer scale and the interaction of nanometer scale objects with light is termed as nanophotonics. The major focus areas of nanophotonics are photonic crystal and surface plasma. This chapter closely examines these key concepts and areas of nanophotonics to provide an extensive understanding of the subject.

Nanophotonics refers to the use of light in nanoscale projects. This field is associated with some specific breakthroughs in using light in new technologies, including silicon-based semiconductors, where nanophotonics improve speed and performance. Nanophotonics is also known as nano-optics.

In this case, nanophotonics involves silicon chips that use light instead of, or in addition to, the types of traditional electrical signals common to semiconductor design. Companies like IBM have pioneered advancements in a chip that uses photodetectors and emitted light to send signals in an integrated circuit environment.

The nanophotonics concept further contributes to a more general category of nanotechnology that is revolutionizing how some of the tiniest projects are treated by the research and development (R&D) departments of various fields.

While nanotechnology has quite a bit of promise, concerns about the uses of nanoscale technologies include the potential rearrangement or disturbance of molecular structures and the effect of nanoscale materials on larger scale environments.

NEAR-FIELD SCANNING OPTICAL MICROSCOPE

Near-field scanning optical microscopy (NSOM) or scanning near-field optical microscopy (SNOM) is a microscopy technique for nanostructure investigation that breaks the far field resolution limit by exploiting the properties of evanescent waves. In SNOM, the excitation laser light is focused through an aperture with a diameter smaller than the excitation wavelength, resulting in an evanescent field (or near-field) on the far side of the aperture. When the sample is scanned at a small distance below the aperture, the optical resolution of transmitted or reflected light is limited only by the diameter of the aperture. In particular, lateral resolution of 20 nm and vertical resolution of 2–5 nm have been demonstrated.

As in optical microscopy, the contrast mechanism can be easily adapted to study different properties, such as refractive index, chemical structure and local stress. Dynamic properties can also be studied at a sub-wavelength scale using this technique.

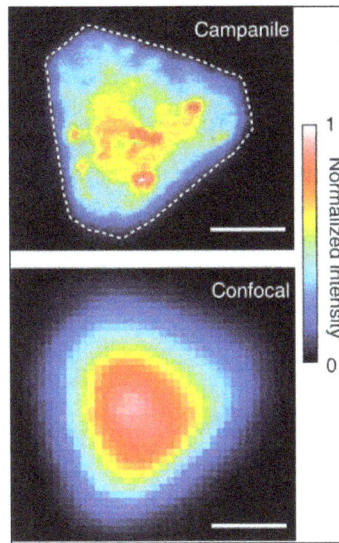

Comparison of photoluminescence maps recorded from a molybdenum disulfide flake using NSOM with a campanile probe (top) and conventional confocal microscopy (bottom). Scale bars: 1 μm.

NSOM/SNOM is a form of scanning probe microscopy.

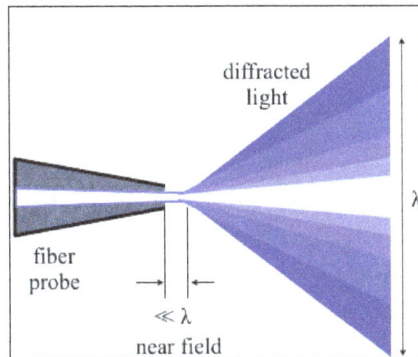

Diagram illustrating near-field optics, with the diffraction of light coming from NSOM fiber probe, showing wavelength of light and the near-field.

Theory

According to Abbe's theory of image formation, developed in 1873, the resolving capability of an optical component is ultimately limited by the spreading out of each image point due to diffraction. Unless the aperture of the optical component is large enough to collect all the diffracted light, the finer aspects of the image will not correspond exactly to the object. The minimum resolution (d) for the optical component are thus limited by its aperture size, and expressed by the Rayleigh criterion:

$$d = 0.61 \frac{\lambda_0}{NA}$$

Here, λ_0 is the wavelength in vacuum; NA is the numerical aperture for the optical component (maximum 1.3–1.4 for modern objectives with a very high magnification factor). Thus, the resolution limit is usually around $\lambda_0/2$ for conventional optical microscopy.

This treatment only assumes the light diffracted into the far-field that propagates without any restrictions. NSOM makes use of evanescent or non propagating fields that exist only near the surface of the object. These fields carry the high frequency spatial information about the object and have intensities that drop off exponentially with distance from the object. Because of this, the detector must be placed very close to the sample in the near field zone, typically a few nanometers. As a result, near field microscopy remains primarily a surface inspection technique. The detector is then rastered across the sample using a piezoelectric stage. The scanning can either be done at a constant height or with regulated height by using a feedback mechanism.

Modes of Operation

Aperture and Apertureless Operation

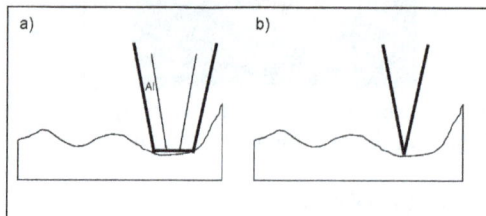

Sketch of a) typical metal-coated tip, and b) sharp uncoated tip.

There exist NSOM which can be operated in so-called aperture mode and NSOM for operation in a non-aperture mode. As illustrated, the tips used in the apertureless mode are very sharp and do not have a metal coating.

Though there are many issues associated with the apertured tips (heating, artifacts, contrast, sensitivity, topology and interference amongst others), aperture mode remains more popular. This is primarily because apertureless mode is even more complex to set up and operate, and is not understood as well. There are five primary modes of apertured NSOM operation and four primary modes of apertureless NSOM operation.

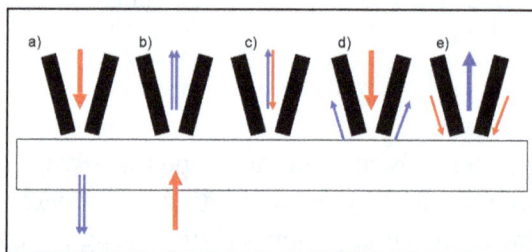

Apertured modes of operation: a) illumination, b) collection, c) illumination collection, d) reflection and e) reflection collection.

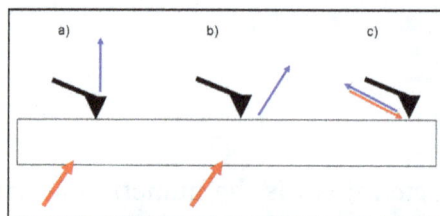

Apertureless modes of operation: a) photon tunneling (PSTM) by a sharp transparent tip, b) PSTM by sharp opaque tip on smooth surface, and c) scanning interferometric apertureless microscopy with double modulation.

Some types of NSOM operation utilize a campanile probe, which has a square pyramid shape with two facets coated with a metal. Such a probe has a high signal collection efficiency (>90%) and no frequency cutoff. Another alternative is "active tip" schemes, where the tip is functionalized with active light sources such as a fluorescent dye or even a light emitting diode that enables fluorescence excitation.

Feedback Mechanisms

Feedback mechanisms are usually used to achieve high resolution and artifact free images since the tip must be positioned within a few nanometers of the surfaces. Some of these mechanisms are constant force feedback and shear force feedback.

Constant force feedback mode is similar to the feedback mechanism used in atomic force microscopy (AFM). Experiments can be performed in contact, intermittent contact, and non-contact modes.

In shear force feedback mode, a tuning fork is mounted alongside the tip and made to oscillate at its resonance frequency. The amplitude is closely related to the tip-surface distance, and thus used as a feedback mechanism.

Contrast

It is possible to take advantage of the various contrast techniques available to optical microscopy through NSOM but with much higher resolution. By using the change in the polarization of light or the intensity of the light as a function of the incident wavelength, it is possible to make use of contrast enhancing techniques such as staining, fluorescence, phase contrast and differential interference contrast. It is also possible to provide contrast using the change in refractive index, reflectivity, local stress and magnetic properties amongst others.

Instrumentation and Standard Setup

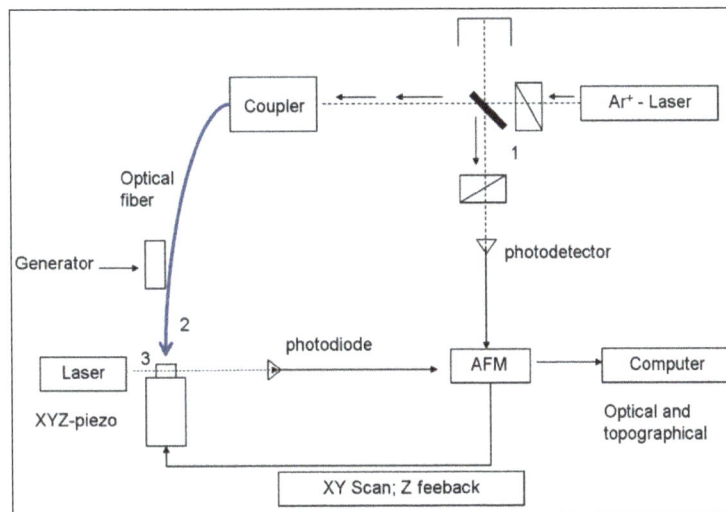

Block diagram of an apertureless reflection-back-to-the-fiber NSOM setup with shear-force distance control and cross-polarization; 1: beam splitter and crossed polarizers; 2: shear-force arrangement; 3: sample mount on a piezo stage.

The primary components of an NSOM setup are the light source, feedback mechanism, the scanning tip, the detector and the piezoelectric sample stage. The light source is usually a laser focused into an optical fiber through a polarizer, a beam splitter and a coupler. The polarizer and the beam splitter would serve to remove stray light from the returning reflected light. The scanning tip, depending upon the operation mode, is usually a pulled or stretched optical fiber coated with metal except at the tip or just a standard AFM cantilever with a hole in the center of the pyramidal tip. Standard optical detectors, such as avalanche photodiode, photomultiplier tube (PMT) or CCD, can be used. Highly specialized NSOM techniques, Raman NSOM for example, have much more stringent detector requirements.

Near-field Spectroscopy

As the name implies, information is collected by spectroscopic means instead of imaging in the near field regime. Through Near Field Spectroscopy (NFS), one can probe spectroscopically with subwavelength resolution. Raman SNOM and fluorescence SNOM are two of the most popular NFS techniques as they allow for the identification of nanosized features with chemical contrast. Some of the common near-field spectroscopic techniques are below.

Direct local Raman NSOM is based on Raman spectroscopy. Aperture Raman NSOM is limited by very hot and blunt tips, and by long collection times. However, apertureless NSOM can be used to achieve high Raman scattering efficiency factors (around 40). Topological artifacts make it hard to implement this technique for rough surfaces.

Tip-enhanced Raman spectroscopy (TERS) is an offshoot of surface enhanced Raman spectroscopy (SERS). This technique can be used in an apertureless shear-force NSOM setup, or by using an AFM tip coated with gold or silver. The Raman signal is found to be significantly enhanced under the AFM tip. This technique has been used to give local variations in the Raman spectra under a single-walled nanotube. A highly sensitive optoacoustic spectrometer must be used for the detection of the Raman signal.

Fluorescence NSOM is a highly popular and sensitive technique makes use of the fluorescence for near field imaging, and is especially suited for biological applications. The technique of choice here is the apertureless back to the fiber emission in constant shear force mode. This technique uses merocyanine based dyes embedded in an appropriate resin. Edge filters are used for removal of all primary laser light.

Near field infrared spectrometry and near field dielectric microscopy use near-field probes to combine sub-micron microscopy with localized IR spectroscopy.

The nano-FTIR method is a broadband nanoscale spectroscopy that uses broadband illumination and FTIR detection to obtain a complete infrared spectrum at every spatial location. Sensitivity to a single molecular complex and nanoscale resolution up to 10 nm has been demonstrated with nano-FTIR.

Artifacts

NSOM can be vulnerable to artifacts that are not from the intended contrast mode. The most common root for artifacts in NSOM are tip breakage during scanning, striped contrast, displaced optical contrast, local far field light concentration, and topographic artifacts.

In apertureless NSOM, also known as scattering-type SNOM or s-SNOM, many of these artifacts are eliminated or can be avoided by proper technique application.

Limitations

One limitation is a very low working distance and extremely shallow depth of field. It is normally limited to surface studies; however, it can be applied for subsurface investigations within the corresponding depth of field. In shear force mode and other contact operation it is not conducive for studying soft materials. It has long scan times for large sample areas for high resolution imaging.

PHOTONIC CRYSTAL

A photonic crystal is a periodic optical nanostructure that affects the motion of photons in much the same way that ionic lattices affect electrons in solids. Photonic crystals occur in nature in the form of structural coloration and animal reflectors, and, in different forms, promise to be useful in a range of applications.

In 1887 the English physicist Lord Rayleigh experimented with periodic multi-layer dielectric stacks, showing they had a photonic band-gap in one dimension. Research interest grew with work in 1987 by Eli Yablonovitch and Sajeev John on periodic optical structures with more than one dimension—now called photonic crystals.

Photonic crystals can be fabricated for one, two, or three dimensions. One-dimensional photonic crystals can be made of layers deposited or stuck together. Two-dimensional ones can be made by photolithography, or by drilling holes in a suitable substrate. Fabrication methods for three-dimensional ones include drilling under different angles, stacking multiple 2-D layers on top of each other, direct laser writing, or, for example, instigating self-assembly of spheres in a matrix and dissolving the spheres.

Photonic crystals can, in principle, find uses wherever light must be manipulated. Existing applications include thin-film optics with coatings for lenses. Two-dimensional photonic-crystal fibers are used in nonlinear devices and to guide exotic wavelengths. Three-dimensional crystals may one day be used in optical computers. Three-dimensional photonic crystals could lead to more efficient photovoltaic cells as a source of power for electronics, thus cutting down the need for an electrical input for power.

Photonic crystals are composed of periodic dielectric, metallo-dielectric—or even superconductor microstructures or nanostructures that affect electromagnetic wave propagation in the same way that the periodic potential in a semiconductor crystal affects electron motion by defining allowed and forbidden electronic energy bands. Photonic crystals contain regularly repeating regions of high and low dielectric constant. Photons (behaving as waves) either propagate through this structure or not, depending on their wavelength. Wavelengths that propagate are called *modes*, and groups of allowed modes form bands. Disallowed bands of wavelengths are called *photonic band gaps*. This gives rise to distinct optical phenomena, such as inhibition of spontaneous emission, high-reflecting omni-directional mirrors, and low-loss-waveguiding. Intuitively, the bandgap of photonic crystals can be understood to arise from the destructive interference of multiple reflections of light propagating in the crystal at the interfaces of the high- and low- dielectric constant regions, akin to the bandgaps of electrons in solids.

The periodicity of the photonic crystal structure must be around half the wavelength of the electromagnetic waves to be diffracted. This is ~350 nm (blue) to ~650 nm (red) for photonic crystals that operate in the visible part of the spectrum—or even less, depending on average index of refraction. The repeating regions of high and low dielectric constant must, therefore, be fabricated at this scale, which is difficult.

Construction Strategies

The fabrication method depends on the number of dimensions that the photonic bandgap must exist in.

Examples of possible photonic crystal structures in 1, 2 and 3 dimensions:

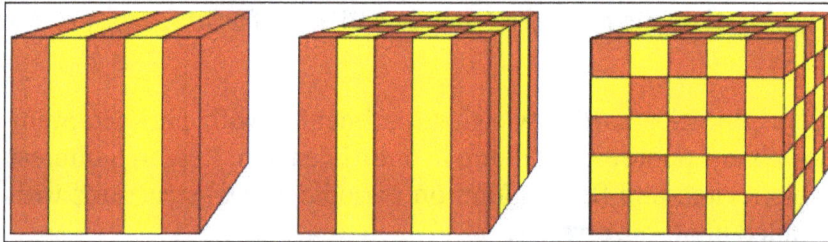

Comparison of 1D, 2D and 3D photonic crystal structures (from left to right, respectively).

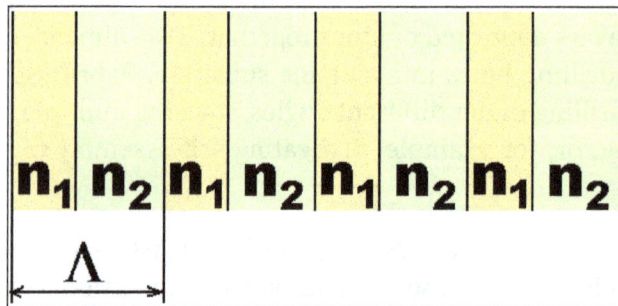

Schematic of a 1D photonic crystal structure, made of alternating layers of a high-dielectric constant material and a low-dielectric constant material. These layers are typically quarter wavelength in thickness.

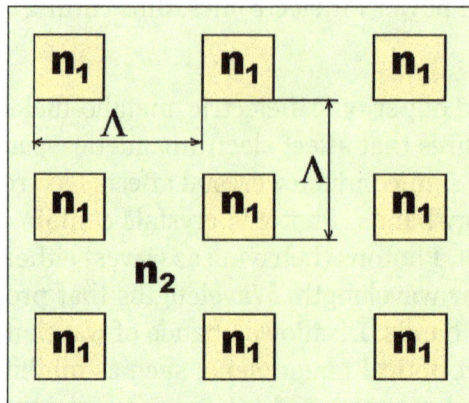

2D photonic crystal structure in a square array.

Schematic of a 2D photonic crystal made of circular holes.

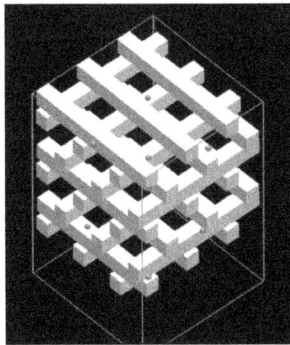

A woodpile structured 3D photonic crystal. These structures have a three-dimensional bandgap for all polarizations.

One-dimensional Photonic Crystals

In a one-dimensional photonic crystal, layers of different dielectric constant may be deposited or adhered together to form a band gap in a single direction. A Bragg grating is an example of this type of photonic crystal. One-dimensional photonic crystals can be either isotropic or anisotropic, with the latter having potential use as an optical switch.

One-dimensional photonic crystal can form as an infinite number of parallel alternating layers filled with a metamaterial and vacuum. This produces identical PBG structures for TE and TM modes.

Recently, researchers fabricated a graphene-based Bragg grating (one-dimensional photonic crystal) and demonstrated that it supports excitation of surface electromagnetic waves in the periodic structure by using 633 nm He-Ne laser as the light source. Besides, a novel type of one-dimensional graphene-dielectric photonic crystal has also been proposed. This structure can act as a far-IR filter and can support low-loss surface plasmons for waveguide and sensing applications. 1D photonic crystals doped with bio-active metals (i.e. silver) have been also proposed as sensing devices for bacterial contaminants.

Two-dimensional Photonic Crystals

In two dimensions, holes may be drilled in a substrate that is transparent to the wavelength of radiation that the bandgap is designed to block. Triangular and square lattices of holes have been successfully employed.

The Holey fiber or photonic crystal fiber can be made by taking cylindrical rods of glass in hexagonal lattice, and then heating and stretching them, the triangle-like airgaps between the glass rods become the holes that confine the modes.

Three-dimensional Photonic Crystals

There are several structure types that have been constructed:

- Spheres in a diamond lattice.

- Yablonovite.

- The woodpile structure – "rods" are repeatedly etched with beam lithography, filled in, and covered with a layer of new material. As the process repeats, the channels etched in each layer are perpendicular to the layer below, and parallel to and out of phase with the channels two layers below. The process repeats until the structure is of the desired height. The fill-in material is then dissolved using an agent that dissolves the fill-in material but not the deposition material. It is generally hard to introduce defects into this structure.

- Inverse opals or Inverse Colloidal Crystals-Spheres (such as polystyrene or silicon dioxide) can be allowed to deposit into a cubic close packed lattice suspended in a solvent. Then a hardener is introduced that makes a transparent solid out of the volume occupied by the solvent. The spheres are then dissolved with an acid such as Hydrochloric acid. The colloids can be either spherical or nonspherical. contains in excess of 750,000 polymer nanorods. Light focused on this beam splitter penetrates or is reflected, depending on polarization.

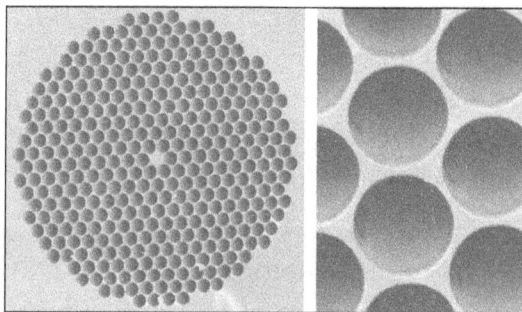

A photonic crystal fiber. SEM images of US NRL-produced fiber. (left) The diameter of the solid core at the center of the fiber is 5 µm, while (right) the diameter of the holes is 4 µm.

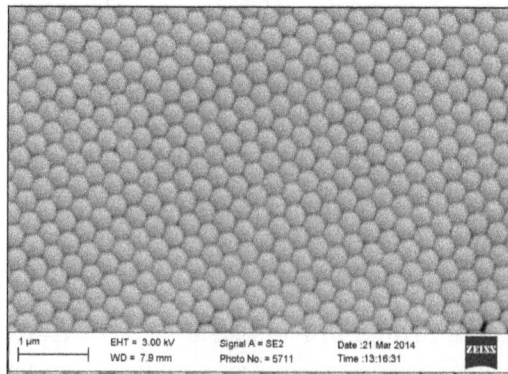

An SEM image of a self-assembled PMMA photonic crystal in two dimensions.

Fabrication Challenges

Higher-dimensional photonic crystal fabrication faces two major challenges:

- Making them with enough precision to prevent scattering losses blurring the crystal properties.

- Designing processes that can robustly mass-produce the crystals.

One promising fabrication method for two-dimensionally periodic photonic crystals is a photonic-crystal fiber, such as a *holey fiber*. Using fiber draw techniques developed for communications fiber it meets these two requirements, and photonic crystal fibres are commercially available. Another promising method for developing two-dimensional photonic crystals is the so-called photonic crystal slab. These structures consist of a slab of material—such as silicon—that can be patterned using techniques from the semiconductor industry. Such chips offer the potential to combine photonic processing with electronic processing on a single chip.

For three dimensional photonic crystals, various techniques have been used—including photolithography and etching techniques similar to those used for integrated circuits. Some of these techniques are already commercially available. To avoid the complex machinery of nanotechnological methods, some alternate approaches involve growing photonic crystals from colloidal crystals as self-assembled structures.

Mass-scale 3D photonic crystal films and fibres can now be produced using a shear-assembly technique that stacks 200–300 nm colloidal polymer spheres into perfect films of fcc lattice. Because the particles have a softer transparent rubber coating, the films can be stretched and molded, tuning the photonic bandgaps and producing striking structural color effects.

Computing Photonic Band Structure

The photonic band gap (PBG) is essentially the gap between the air-line and the dielectric-line in the dispersion relation of the PBG system. To design photonic crystal systems, it is essential to engineer the location and size of the bandgap by computational modeling using any of the following methods:

- Plane wave expansion method.

- Finite element method.

- Finite difference time domain method.

- Order-n spectral method.

- KKR method.

- Bloch wave – MoM method.

- Construction of the Band Diagram.

Essentially, these methods solve for the frequencies (normal modes) of the photonic crystal for each value of the propagation direction given by the wave vector, or vice versa. The various lines in the band structure, correspond to the different cases of n, the band index.

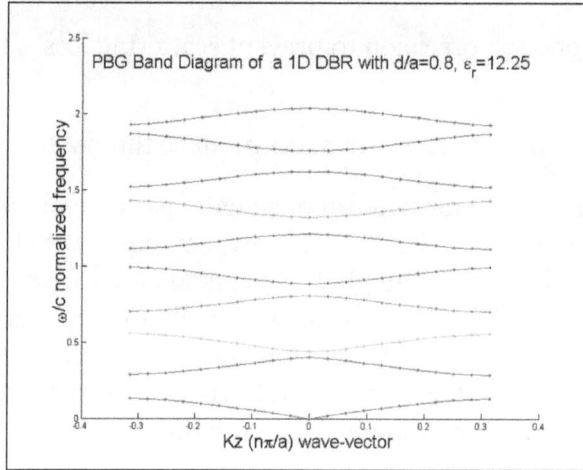

Band structure of a 1D photonic crystal, DBR air-core calculated using plane wave expansion technique with 101 planewaves, for d/a=0.8, and dielectric contrast of 12.250.

The plane wave expansion method can be used to calculate the band structure using an eigen formulation of the Maxwell's equations, and thus solving for the eigen frequencies for each of the propagation directions, of the wave vectors. It directly solves for the dispersion diagram. Electric field strength values can also be calculated over the spatial domain of the problem using the eigen vectors of the same problem. For the picture shown to the right, corresponds to the band-structure of a 1D distributed Bragg reflector (DBR) with air-core interleaved with a dielectric material of relative permittivity 12.25, and a lattice period to air-core thickness ratio (d/a) of 0.8, is solved using 101 planewaves over the first irreducible Brillouin zone.

To speed calculation of the frequency band structure, the Reduced Bloch Mode Expansion (RBME) method can be used. The RBME method applies "on top" of any of the primary expansion methods mentioned above. For large unit cell models, the RBME method can reduce time for computing the band structure by up to two orders of magnitude.

Applications

Photonic crystals are attractive optical materials for controlling and manipulating light flow. One dimensional photonic crystals are already in widespread use, in the form of thin-film optics, with applications from low and high reflection coatings on lenses and mirrors to colour changing paints and inks. Higher-dimensional photonic crystals are of great interest for both fundamental and applied research, and the two dimensional ones are beginning to find commercial applications.

The first commercial products involving two-dimensionally periodic photonic crystals are already available in the form of photonic-crystal fibers, which use a microscale structure to confine light with radically different characteristics compared to conventional optical fiber for applications in nonlinear devices and guiding exotic wavelengths. The three-dimensional counterparts are still far from commercialization but may offer additional features such as optical nonlinearity required for the operation of optical transistors used in optical computers, when some technological aspects such as manufacturability and principal difficulties such as disorder are under control.

SURFACE PLASMON

Surface plasmons (SPs) are coherent delocalized electron oscillations that exist at the interface between any two materials where the real part of the dielectric function changes sign across the interface (e.g. a metal-dielectric interface, such as a metal sheet in air). SPs have lower energy than bulk (or volume) plasmons which quantise the longitudinal electron oscillations about positive ion cores within the bulk of an electron gas (or plasma).

The charge motion in a surface plasmon always creates electromagnetic fields outside (as well as inside) the metal. The *total* excitation, including both the charge motion and associated electromagnetic field, is called either a surface plasmon polariton at a planar interface, or a localized surface plasmon for the closed surface of a small particle.

The existence of surface plasmons was first predicted in 1957 by Rufus Ritchie. In the following two decades, surface plasmons were extensively studied by many scientists, the foremost of whom were T. Turbadar in the 1950s and 1960s, and Heinz Raether, E. Kretschmann, and A. Otto in the 1960s and 1970s. Information transfer in nanoscale structures, similar to photonics, by means of surface plasmons, is referred to as plasmonics.

Surface Plasmon Polaritons

Excitation

Surface plasmon polaritons can be excited by electrons or photons. In the case of photons, it cannot be done directly, but requires a prism, or a grating, or a defect on the metal surface.

Dispersion Relation

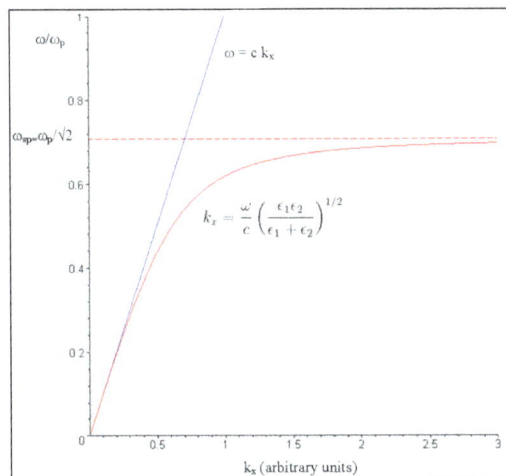

Dispersion curve for surface plasmons. At low k, the surface plasmon curve (red) approaches the photon curve (blue).

At low frequency, an SPP approaches a Sommerfeld-Zenneck wave, where the dispersion relation (relation between frequency and wavevector) is the same as in free space. At a higher frequency, the dispersion relation bends over and reaches an asymptotic limit called the "surface plasma frequency".

Propagation Length and Skin Depth

As an SPP propagates along the surface, it loses energy to the metal due to absorption. It can also lose energy due to scattering into free-space or into other directions. The electric field falls off evanescently perpendicular to the metal surface. At low frequencies, the SPP penetration depth into the metal is commonly approximated using the skin depth formula. In the dielectric, the field will fall off far more slowly. SPPs are very sensitive to slight perturbations within the skin depth and because of this, SPPs are often used to probe inhomogeneities of a surface.

Experimental Applications

The excitation of surface plasmons is frequently used in an experimental technique known as surface plasmon resonance (SPR). In SPR, the maximum excitation of surface plasmons are detected by monitoring the reflected power from a prism coupler as a function of incident angle or wavelength. This technique can be used to observe nanometer changes in thickness, density fluctuations, or molecular absorption.

Surface plasmon-based circuits have been proposed as a means of overcoming the size limitations of photonic circuits for use in high performance data processing nano devices.

The ability to dynamically control the plasmonic properties of materials in these nano-devices is key to their development. A new approach that uses plasmon-plasmon interactions has been demonstrated recently. Here the bulk plasmon resonance is induced or suppressed to manipulate the propagation of light. This approach has been shown to have a high potential for nanoscale light manipulation and the development of a fully CMOS- compatible electro-optical plasmonic modulator, said to be a future key component in chip-scale photonic circuits.

Some other surface effects such as surface-enhanced Raman scattering and surface-enhanced fluorescence are induced by surface plasmon of noble metals, therefore sensors based on surface plasmons were developed.

In surface second harmonic generation, the second harmonic signal is proportional to the square of the electric field. The electric field is stronger at the interface because of the surface plasmon resulting in a non-linear optical effect. This larger signal is often exploited to produce a stronger second harmonic signal.

The wavelength and intensity of the plasmon-related absorption and emission peaks are affected by molecular adsorption that can be used in molecular sensors. For example, a fully operational prototype device detecting casein in milk has been fabricated. The device is based on monitoring changes in plasmon-related absorption of light by a gold layer.

SINGLE-PHOTON SOURCE

Single-photon sources are light sources that emit light as single particles or photons. They are distinct from coherent light sources (lasers) and thermal light sources such as incandescent light bulbs. The Heisenberg uncertainty principle dictates that a state with an exact number of photons of a single frequency cannot be created. However, Fock states (or number states) can be studied for a system where the electric field amplitude is distributed over a narrow bandwidth. In this context, a single-photon source gives rise to an effectively one-photon number state. Photons from an ideal single-photon source exhibit quantum mechanical characteristics. These characteristics include photon antibunching, so that the time between two successive photons is never less than some minimum value. This is normally demonstrated by using a beam splitter to direct about half of the incident photons toward one avalanche photodiode, and half toward a second. Pulses from one detector are used to provide a 'counter start' signal, to a fast electronic timer, and the other, delayed by a known no. of nanoseconds, is used to provide a 'counter stop' signal. By repeatedly measuring the times between 'start' and 'stop' signals, one can form a histogram of time delay between two photons and the coincidence count- if bunching is not occurring, and photons are indeed well spaced, a clear notch around zero delay is visible.

In quantum theory, photons describe quantized electromagnetic radiation. Specifically, a photon is an elementary excitation of a normal mode of the electromagnetic field. Thus a single-photon state is the quantum state of a radiation mode that contains a single excitation.

Single radiation modes are labelled by, among other quantities, the frequency of the electromagnetic radiation that they describe. However, in quantum optics, single-photon states also refer to mathematical superpositions of single-frequency (monochromatic) radiation modes. This definition is general enough to include photon wave-packets, i.e., states of radiation that are localized to some extent in space and time.

Single-photon sources generate single-photon states as described above. In other words, ideal single-photon sources generate radiation with a photon-number distribution that has a mean one and variance zero.

Characteristics

An ideal-single photon source produces single-photon states with 100% probability and optical vacuum or multi-photon states with 0% probability. Desirable properties of real-world single-photon sources include efficiency, robustness, ease of implementation and on-demand nature, i.e., generating single-photons at arbitrarily chosen times. Single-photon sources including single emitters such as single atoms, ions and molecules, and including solid-state emitters such as quantum dots, color centers and carbon nanotubes are on-demand. Currently, there are many active nanomaterials engineered into single quantum emitters where their spontaneous emission could be tuned by changing the local density of optical states in dielectric nanostructures. The dielectric nanostructures are usually designed within the heterostructures to enhance the light-matter interaction, and thus further improve the efficiency of these single photon sources. Another type of source comprises non-deterministic sources, i.e., not on demand, and these include examples such as weak lasers, atomic cascades and parametric down-conversion.

The single-photon nature of a source can be quantized using the second-order correlation function $g^{(2)}(\tau)$. Ideal single-photon sources show $g^{(2)}(0) = 0$ and good single-photon sources have small $g^{(2)}(0)$. The second-order correlation function can be measured using the Hanbury-Brown–Twiss effect.

Types

The generation of a single photon occurs when a source creates only one photon within its fluorescence lifetime after being optically or electrically excited. An ideal single-photon source has yet to be created. Given that the main applications for a high-quality single-photon source are quantum key distribution, quantum repeaters and quantum information science, the photons generated should also have a wavelength that would give low loss and attenuation when travelling through an optical fiber. Nowadays the most common sources of single photons are single molecules, Rydberg atoms, diamond colour centres and quantum dots, with the last being widely studied with efforts from many research groups to realize quantum dots that fluoresce single photons at room temperature with photons in the low loss window of fiber-optic communication. For many purposes single photons need to be anti-bunched, and this can be verified.

Faint Laser

One of the first and easiest sources was created by attenuating a conventional laser beam to reduce its intensity and thereby the mean photon number per pulse. Since the photon statistics follow a Poisson distribution one can achieve sources with a well defined probability ratio for the emission of one versus two or more photons. For example a mean value of $\mu = 0.1$ leads to a probability of 90% for zero photons, 9% for one photon and 1% for more than one photon.

Although such a source can be used for certain applications, it has a second-order intensity correlation function equal to one (no antibunching). For many applications however, antibunching is required, for instance in quantum cryptography.

Heralded Single Photons

Pairs of single photons can be generated in highly correlated states from using a single high-energy photon to create two lower-energy ones. One photon from the resulting pair may be detected to 'herald' the other (so its state is pretty well known prior to detection). The two photons need not generally be the same wavelength, but the total energy and resulting polarisation are defined by the generation process. One area of keen interest for such pairs of photons is QKD.

The heralded single-photon sources are also used to examine the fundamental physics laws in quantum mechanics. There are two commonly used types of heralded single-photon sources: spontaneous parametric down-conversion and spontaneous four-wave mixing. The first source has line-width around THz and the second one has line-width around MHz or narrower. The heralded single photon has been used to demonstrate photonics storage and loading to the optical cavity.

Probabilistic Single-photon Sources

Spontaneous Parametric Down-conversion (SPDC)

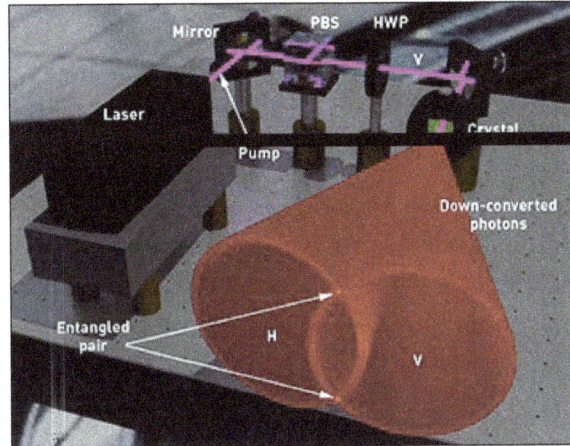

In figure, type-II SPDC. In spontaneous parametric down-conversion (SPDC), a nonlinear medium (for example, a birefringent crystal), illuminated by a pump laser, facilitates a process in which a pump photon is absorbed to generate two correlated daughter photons, the signal and the idler. In so-called type-II SPDC, the signal and idler photons have perpendicular polarizations and emerge as two cones—with entangled photons found where the cones intersect.

SPDC is the most popular means of generating single photons. In this second-order nonlinear optical process, a high-energy pump photon within a nonlinear optical medium generates two lower-energy photons, the "signal" and "idler" photons. As the name implies, the process is spontaneous (generated by quantum vacuum fields), parametric (the initial and final quantum-mechanical states of the medium are identical, and photon energy is always conserved), and an example of down-conversion (as the signal and idler frequencies are always lower than that of the pump).

The SPDC process obeys energy and momentum conservation, which are together known as the phase-matching conditions. SPDC can be used in so-called heralded schemes to generate single photons. As both photons are generated at the same time, detecting one heralds the presence of the other; this stream of heralded photons can then be used for single-photon applications.

SPDC has a number of key advantages. While the process efficiency is low, with only 1 in 10^9 or 10^{10} pump photons undergoing SPDC, the process still accounts for fairly bright photon sources with a wide range of applications, generating photons at a rate of about 2 MHz. In addition, SPDC excels in creating entangled photons; more than 99.5% fidelity of entanglement has been measured in entangled photons produced by the process. As a result, SPDC has found numerous applications in foundational tests of quantum mechanics, quantum information processing, quantum metrology and quantum communication.

One possible downside of SPDC—beyond its probabilistic nature, which makes it unsuitable for applications requiring photons on demand—is that SPDC carries a finite probability of multiphoton events. These can cause security loopholes in QKD protocols, although the situation is much better than with WCP sources, for which such probabilities are much higher.

Four-wave Mixing

In another scheme for single-photon generation, four-wave mixing, two pump photons convert to a signal and an idler photon in the presence of a third-order nonlinear medium. Like SPDC, this requires satisfaction of phase-matching conditions. As opposed to SPDC, however, which can occur both in bulk and in confined geometries such as cavities and waveguides (which, in turn, narrows down the emission direction), four-wave mixing has generally been demonstrated only in integrated-optics structures involving waveguides.

Waveguide-based sources generally can have higher pair generation rates than bulk sources, due to a lower number of interacting modes. Four-wave-mixing-based photon sources have reported brightnesses of around 0.855 MHz, with a current record for entanglement fidelity of 99.7%.

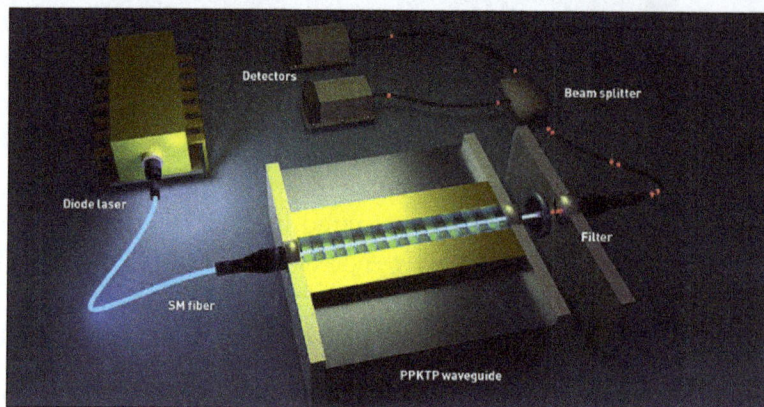

In figure, single photons in fiber: In this setup, a continuous-wave pump beam is coupled to a single-mode fiber. This fiber is aligned to a waveguide of periodically poled potassium titanyl phosphate (PPKTP), a popular nonlinear material. SPDC photon pairs (red bullets) are generated continuously and passed through a filter, which blocks any pump light. Photons whose polarizations are orthogonal to each other can be separated by a polarizing beam splitter and detected in two single-photon avalanche detectors.

On-demand Single-photon Sources

Atom and Ion Sources

Among the simplest and most elegant architectures for single-photon sources involve photon emission due to transitions in atoms or ions. The first so-called single-photon source was an atomic-cascade-based source, and such sources were used in a number seminal experiments, such as demonstration of violation of the Bell inequality as well as double-slit interference using photons. The first antibunching experiment was successfully observed in resonant fluorescence of sodium atoms.

A single excited atom spontaneously emits a single photon. But isolating a single atom was historically a challenge. Then, the invention of ion trapping and atom trapping led to the investigation of trapped-ion and trapped-atom-based photon sources. Emission from such sources happens in all directions, which leads to poor detection efficiency. One way to boost the efficiency is to place atoms or ions in a cavity; in such a setup, the emission modes couple to the cavity mode, giving it directionality and thus increasing the efficiency.

An important advantage of trapped-atom and trapped-ion sources is the narrow emission linewidth that they offer. Thus they are very good in interferometric applications. They generate pure single photons, as the probability of multiple photon emission is quite low in general. While not the optimal choice for technological applications, they remain an excellent choice for experiments in quantum foundations as well as in metrology.

Quantum Dots

"Artificial" atoms, or quantum dots, constitute one of the most promising photon sources on the near horizon, with much potential for future development and applications. Quantum dots are tiny semiconductor particles or nanocrystals, with diameters in the 2-to-10-nm range. They are referred to as artificial atoms because of their atom-like discrete energy spectra.

When illuminated by an optical pulse, electrons in the quantum dot and its vicinity jump from the valence to the conduction band, leading to the formation of electron–hole pairs that then rapidly relax to the lowest energy states. Depending on the population, one can observe recombination from the exciton, the bi-exciton and multi-exciton states, labels that refer to the number of electron-hole pairs left to recombine. Although all such transitions possess distinct energies and can, in principle, serve as emitters for single photons by spectral filtering, the bi-excitonic and excitonic transitions are the most commonly used.

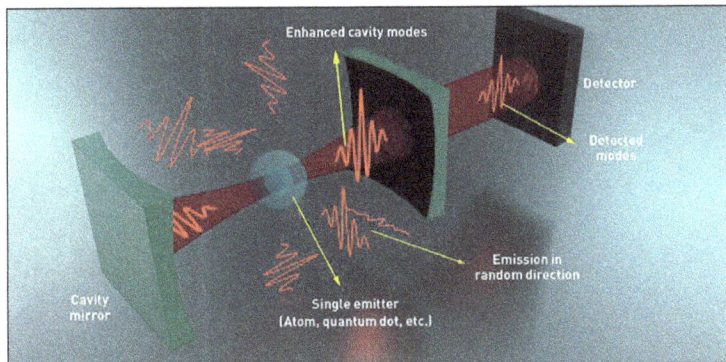

In figure, atoms in cavities: Both real atoms and "artificial atoms" (quantum dots), when excited, undergo spontaneous emission of photons in all directions. Atomic cascades and bi-exciton modes in quantum dots have short-lived metastable states and thus emit photons in pairs. Placing the atom or quantum dot in a cavity allows the emission modes resonant to the cavity to be enhanced, amplifying the emission in a given spatial direction.

Quantum dot sources have also been used to produce highly entangled photon pairs. At present, the highest single-photon emission rate for quantum dots stands at 28.3 MHz, and an entanglement fidelity of 97.8% has been demonstrated. A quantum dot source can be either optically or electrically pumped. Electrical pumping holds the higher brightness record, and electrically pumped sources have even been realized at room temperature. On the other hand, the $g^{(2)}(0)$ for optically pumped quantum dot sources can be considerably lower than for electrically pumped ones—an indication that optical pumping may provide something closer to a true single-photon source.

Self-assembled quantum dots have been integrated with cavities to increase the directionality of emission and ease of use. The structural integrity of these cavities continues to improve, thereby enhancing the quality factor. Epitaxial deposition of quantum dots sometimes causes them to grow at random positions on a surface, which leads to losses in collection efficiency.

Continuous improvement is taking place in the quantum dot domain, with brightness and entanglement records being broken frequently. These deterministic sources have been successfully used in various QKD implementations as well as experimental demonstration of multiphoton Boson sampling, among other applications.

References

- Nanophotonics- 29146: techopedia.com, Retrieved 03 June, 2019

- Single-photon-sources: osa-opn.org, Retrieved 28 August, 2019

- Oshikane, Y.; et al. (2007). "Observation of nanostructure by scanning near-field optical microscope with small sphere probe". Sci. Technol. Adv. Mater. (free access)|format= requires |url= (help). 8 (3): 181. Bibcode:2007STAdM...8..181O. doi:10.1016/j.stam.2007.02.013

- John D Joannopoulos; Johnson SG; Winn JN; Meade RD (2008), "Photonic Crystals: Molding the Flow of Light", Photonic Crystals: Molding the Flow of Light (2nd ed.), Bibcode:2008pcmf.book.....J, ISBN 978-0-691-12456-8

PERMISSIONS

We would like to thank the editorial team for lending their expertise to make the book truly unique. They have played a crucial role in the development of this book. Without their invaluable contributions this book wouldn't have been possible. They have made vital efforts to compile up to date information on the varied aspects of this subject to make this book a valuable addition to the collection of many professionals and students.

This book was conceptualized with the vision of imparting up-to-date and integrated information in this field. To ensure the same, a matchless editorial board was set up. Every individual on the board went through rigorous rounds of assessment to prove their worth. After which they invested a large part of their time researching and compiling the most relevant data for our readers.

The editorial board has been involved in producing this book since its inception. They have spent rigorous hours researching and exploring the diverse topics which have resulted in the successful publishing of this book. They have passed on their knowledge of decades through this book. To expedite this challenging task, the publisher supported the team at every step. A small team of assistant editors was also appointed to further simplify the editing procedure and attain best results for the readers.

Apart from the editorial board, the designing team has also invested a significant amount of their time in understanding the subject and creating the most relevant covers. They scrutinized every image to scout for the most suitable representation of the subject and create an appropriate cover for the book.

The publishing team has been an ardent support to the editorial, designing and production team. Their endless efforts to recruit the best for this project, has resulted in the accomplishment of this book. They are a veteran in the field of academics and their pool of knowledge is as vast as their experience in printing. Their expertise and guidance has proved useful at every step. Their uncompromising quality standards have made this book an exceptional effort. Their encouragement from time to time has been an inspiration for everyone.

The publisher and the editorial board hope that this book will prove to be a valuable piece of knowledge for students, practitioners and scholars across the globe.

INDEX